The
Kazan
Journey into an Emerging Land

KENSINGTON PALACE

The Arctic and its people remain a distant mystery for most of us although I have been lucky enough in the past to visit various parts of it. Scientists and explorers have been captivated by the challenge of unlocking that mystery for over four hundred years. The environment is different from any other on Earth.

Seven thousand years ago the valley of the Kazan River emerged, as the final glacier of the last ice age retreated. Simultaneously this area of the mainland arctic contains some of the oldest surface rock and is one of the youngest landscapes on our planet. After deglaciation an entire ecosystem developed to the point where it was capable of supporting human life - the early Indians and Eskimos. But the mystery of that evolution and of the human use of its resources remains.

To seek solutions of that mystery is important, both for the land and for its inhabitants. Only in the quest to understand do people develop a respect. The more we understand the unique environment that has emerged in the arctic barrenlands, the more we will value that wilderness. And only in establishing a record of the human life that once depended on that environment, can we ensure the preservation of a cultural heritage unlike any other in the world.

The valley of the Kazan River today is a wilderness. No one lives there and few travel there. The mystery of the emerging land endures. Its natural and cultural heritage is rich. In an awareness of that heritage lies the key to its preservation.

Charles

NORTHERN HERITAGE SERIES:

PEOPLE AND PLACES

SERIES EDITOR:

Christopher Stephens

THE ERA FOUNDATION

This publication has been made possible by the generous support of the United Nations Educational, Scientific and Cultural Organization, the TUNDIA Council and the World Decade for Cultural Development, the Imperial Oil Family of Companies and Esso Resources (Ltd.)

NORTHERN HERITAGE SERIES: PEOPLE AND PLACES

LITERATURE IN THE SERVICE OF THE EARTH

There is a moral dilemma faced by people throughout the world. It is a problem which would not have been admitted, universally, a century ago. How can people and earth serve each other to the benefit of both - guaranteeing survival and evolution? To whom or what should we turn?

It is increasingly difficult to find an answer buried in the complexities of 'first' world dogmas and teachings: the literature, the arts and the cultures too convoluted for simple sourcing. Instead, we find the burgeoning expressions of indigenous peoples - confronted with an accelerating global change - to be clear, genuine and simple solutions.

Nowhere is this more apparent than in the contributions made by circumpolar peoples. Northern traditional knowledge comes from two sources - people and places. The former is referred to as 'foundtruth', the latter as 'groundtruth'. This series shows how the two can provide answers to such things as compassionate education, sustainable environmental development, spiritual economics, accessible heritage conservation and art ecology. Drawing on the work of northerners from nine circumpolar nations, the *Northern Heritage Series: People and Places* is designed to ensure a legacy of wisdom exists for future generations.

Series Editor: Christopher Stephens

The Kazan

Journey into an Emerging Land

edited by

David F. Pelly

Christopher C. Hanks

OUTCROP,

The Northern Publishers, Yellowknife, NWT, Canada

Series design: Heidi Held, Outcrop Ltd.

Cover photos: Mark Côté.

Canadian Cataloguing in Publication Data

Main entry under title:

The Kazan: Journey into an emerging land

(Northern heritage series: People and places)

Includes bibliographical references.

ISBN 0-919315-26-7

1. Operation Raleigh Canada. 2. Kazan River (N.W.T.) - Description and travel. 3. Scientific expeditions - Northwest Territories - Kazan River Valley. 4. Natural history - Northwest Territories - Kazan River Valley. 5. Northwest Territories - Description and travel - 1981 - * I. Pelly, David F. (David Fraser), 1948- II. Hanks, Christopher C. III. Series.
FC4195.K39K39 1991 917.19'2043 C91-091716-7
F1060.K39 1991

Outcrop Ltd.
The Northern Publishers
Box 1350, Yellowknife, Northwest Territories
Canada X1A 2N9

Printed and bound in Canada

Inuit Ku, River of Men,
Highway into the heart of the sky;
Caribou tracks on my soul.

J.K.

TABLE OF CONTENTS

FOREWORD

At a time in history when issues of world order are best settled on the international stage, it is encouraging to know that certain endeavours can succeed entirely due to this kind of global cooperation. In fact, achievements in dealing with global issues such as environmental change, indigenous rights, and cultural plurality are due to those ventures which mix the best of human knowledge with genuine experience in some of the unique natural regions of the world. Such is the success of the Canadian Arctic Expedition.

Operation Raleigh's Canadian Arctic Expedition was conducted on the Kazan River in the Keewatin region of Canada's Northwest Territories in July and August of 1988. Operation Raleigh was a four-year around-the-world-expeditionary venture cast in three month phases. The project coincided with the 400th anniversary of Sir Walter Raleigh's circumnavigation of the globe; its combined purpose was science and community service on an international scale in 40 countries. Participants, between the ages 17-24, were chosen from eleven different countries for the Canadian phase.

From the outset, the Canadian Arctic Expedition spelled partnership. The goal of the expedition was to traverse the Kazan River in the Canadian eastern Arctic, a 500 kilometre canoe expedition taking seven weeks. The intention was to conduct a multidisciplinary scientific study of the natural and cultural heritage. The method matched the environs: the Arctic land included a complexity of relationships existing between climate, animal, bird, and human which in turn demanded that the expedition members coordinate their ecological, biological, geological and archaeological surveys. Prior to the commencement of the river travel, the participants spent a month in scientific orientation and survival training at Moorelands Camp and Trent University, Peterborough in Ontario. While on the river, venturers from 11 different countries, four scientists and four group leaders travelled in four separate groups to enhance the quality and variety of data collected and reduce the amount of actual impact on the land. The alliance created between the science leaders and the young venturers was crucial to the success of the expedition.

There arose with this expedition, almost instinctively, a partnership between human and environment. The bridge between nature and culture was laid through a direct experience adventure. It is one thing to know the Arctic vicariously; it is quite another to have it know you directly. The fine lines of national and cultural distinction as represented by the venturers blurred with the horizons as the expedition progressed. Participants became just another feature, just one more element, in the arctic mosaic. As His Royal Highness the Prince of Wales reminds us: "only in the quest to understand do people develop a respect".

In another sense, this particular expedition defined a new kind of partnership between individual and habitat. As we are reminded by the travel diary entries of the venturers and by the synopses, the arctic ecosystem has a constrained set of opportunities and parameters for adaptation and adjustment to change. It became clear to all those involved that in the words of their chief scientist, Chris Hanks, the storyline had changed for humans - "They were no longer players, but stage managers". And so the role devolved to the students whose learning was not simply as observers but perhaps, upon returning to their countries, as environmental protectors and instigators of change. Yet another realization of this was the fact that the Arctic, like the rain forest, desert, and urban environment were each experiencing similar problems - loss of species, dwindling natural areas, and depletion of resources.

Finally, the logistics of this expedition required a partnership in vision and organization. David Pelly, as leader and coordinator, had no ordinary task. He succeeded because he matched vision with reality. The reality was internationally based and joined through multinational efforts. The vision was ably supported by 77 public and private sponsors, all sharing the same respect for the Arctic and the opportunity it offered for allowing one expeditionary venture to make a significant step in, not neccessarily resolving world issues, but at least addressing the basic requirements for making the world a better place to live. The fact that the Arctic continues to enjoin the work of other international projects - such as 'Icewalk', the Canada-Soviet Polar Ski Trek, and 'Masters of the Arctic' - is a sure sign that the legacy for a future world will draw many of its principles from a circumpolar experience.

We have much to thank the members of the Canadian Arctic Expedition and they, in turn, much to show in appreciation, for their northern heritage experience.

Christopher Stephens
The ERA Foundation

NORTHERN HERITAGE SERIES: PATRON'S WORD

It has been the primary goal of Imperial Oil Limited and its family of companies to foster the development of renewable resources so that people may benefit from their use. Successful resource development depends on a fundamental belief: we conserve that which is limited and preserve the knowledge which allows us to do this. This dictum applies especially in the northern regions of our work.

When we speak of renewable resources we refer not just to the natural environment but also to the traditional knowledge and cultural heritage of people whose lives are inextricably tied to the land. Resource conservation and sustainable development are only possible if there exists an enduring respect for the environment. Such respect is only achieved by working with people whose awareness of the land has been tempered by historical changes. It is these sensitivities which serve as testimonies of principle concerning right and wrong action.

I am especially supportive of the Northern Heritage Series as it will provide people throughout the world with a legacy of knowledge gleaned from millenia and centuries of circumpolar life. It is both timely and fitting that the north, because of its leadership and vision in maintaining environmental quality, should see to the expression of its views in this literary manner. I look forward to learning about the north and its people by way of these books and their subjects of natural and cultural heritage.

Arden R. Haynes
Chairman and Chief Executive Officer
Imperial Oil Limited

INTRODUCTION

The Canadian Arctic Expedition on the Kazan River was a component of the international youth program Operation Raleigh. From 1984-89, this British-based initiative for service to science and community conducted expeditions in forty countries, involving a total of more than 4000 young people from all around the world. His Royal Highness The Prince of Wales was the Patron of Operation Raleigh.

The 24 young participants on the Kazan expedition came from eleven different nations, selected in their countries of origin as suitable representatives who stood to benefit from the experience. They were accompanied by a team of eight staff, scientists and wilderness trip leaders. These 32 travelled in four teams of eight each, to maximize the scientific coverage and minimize environmental impact.

The Kazan flows through the tundra of Canada's mainland Arctic, west of Hudson Bay. It emerges from the spruce forests near the border between Manitoba and the Northwest Territories, and winds northward to its terminus at Baker Lake, just 250 km south of the Arctic Circle.

The scientific projects of the Canadian Arctic Expedition were developed during five years of planning by David Pelly and a committee of volunteers. Canada's scientific community was canvassed for projects that would meet the needs of researchers, wildlife managers, and institutions across Canada, and result in an unprecedented multi-disciplinary survey of an arctic river valley.

The scientific survey concentrated on the ecosystem as it had developed since the last ice age. The expedition examined the soil, the vegetation, the birds and mammals, and the evidence of past human occupation in the Kazan valley. The connection between these living elements became a theme of the expedition, and its legacy.

For each of the participants the expedition was a personal voyage of discovery. Some chose to record their experiences in paintings, sketches, photographs, and journals. Others preferred simply to absorb their thoughts into memory. But for all involved, the Kazan expedition was an important and memorable event in their lives.

ACKNOWLEDGEMENTS

The expedition described in *The Kazan, Journey into an Emerging Land* was made possible by the support of Col. John Blashford-Snell MBE and all those associated with Operation Raleigh.

The scientific projects were willingly supported by the Government of the Northwest Territories. The former Commissioner of the Northwest Territories - John H. Parker; Christopher Stephens and Charles Arnold of the Prince of Wales Northern Heritage Centre; Robert Janes of the Science Institute; and Robert Bromley, Douglas Heard, and Christopher Shank of the Department of Renewable Resources deserve special thanks for their assistance.

Fred Helleiner, of Trent University, believed from the beginning that amateurs could make a substantial contribution to bird atlasing in the Northwest Territories.

Peter Storck, of the Department of New World Archaeology at the Royal Ontario Museum, provided invaluable support to the archaeological research programme.

Marc Côté acted as official photographer for the expedition and many of the photographs in this book are his.

Kassie Heath, whose sketches appear throughout the book, was one of the venturer participants on the trip.

We wish to especially acknowledge the venturers of the Canadian Arctic Expedition who shared their personal journals with us. The success of the scientific research is largely the result of all the venturer's dedicated labours.

Journey into an Emerging Land

Kazan River Region

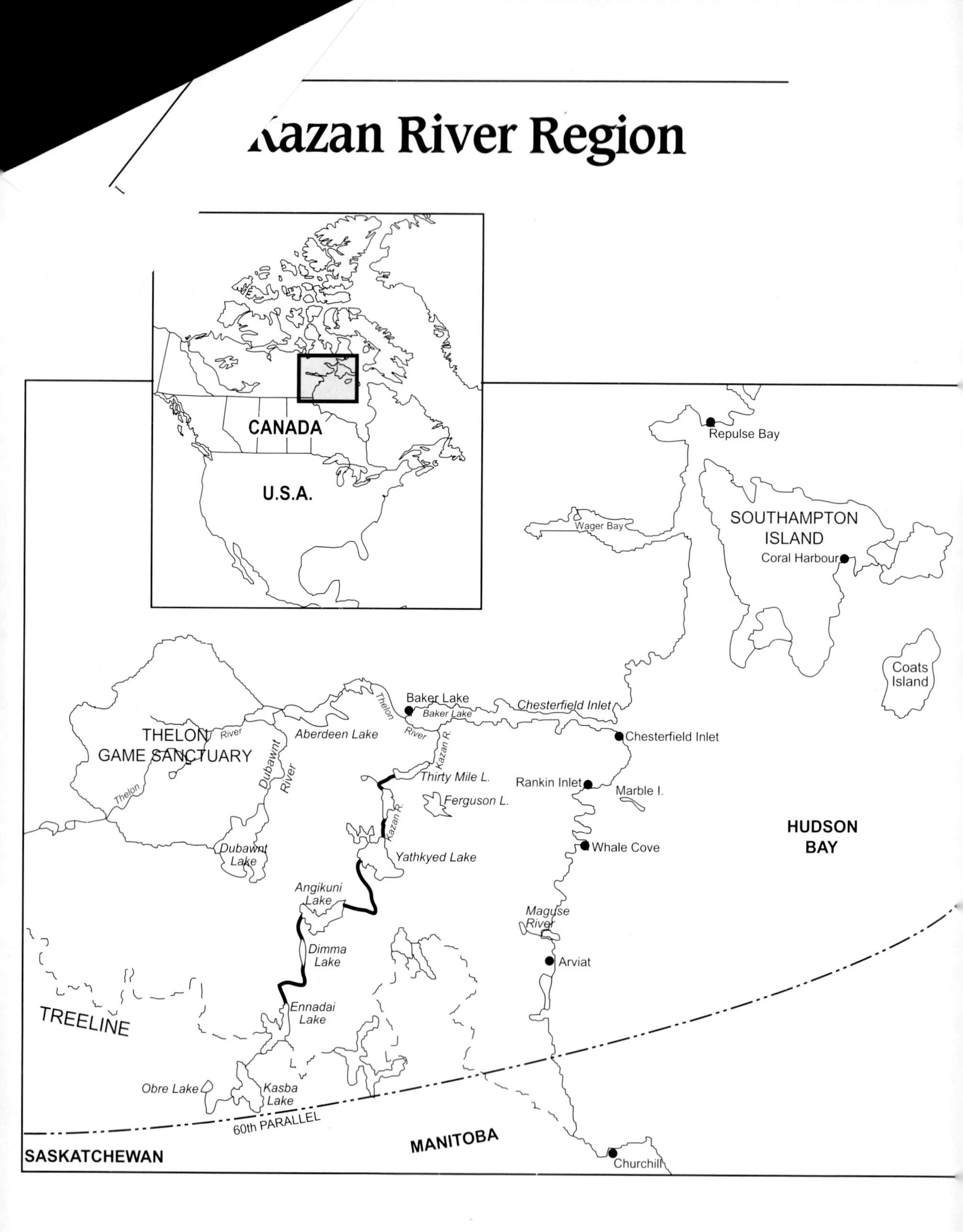

THE JOURNEY BEGINS

North of the border across the top of Manitoba and Saskatchewan sits the largest area of deep wilderness left in North America, perhaps the world. It begins in the boreal forest of the Canadian provinces and extends north into the Northwest Territories, where the Canadian Shield spreads west from Hudson Bay. As you advance north, the trees become smaller and more sparse until eventually most of the land is devoid of any growth higher than a man's waist. Beyond the tree-line the country is known as the barrenlands - a million square kilometres of seemingly endless space, extending north to the Arctic Ocean.

It is a land of gently rolling hills, of sand eskers and granite outcroppings, of lakes, tundra plains, and broad rivers. It is still the home of the once semi-nomadic Caribou Inuit, the "People of the Deer", people who depend on the caribou for many of the material needs of life. It is here that the great herds of caribou, hundreds of thousands strong, cross to and fro in an endless cycle of seasonal migrations.

Three river valleys dominate the map of the barrens - the Thelon, Dubawnt and Kazan. They rise at the outer limits of the barrenlands and follow their independent courses through the wilderness to join together in the very heart of the land, before flowing out to the sea.

The Thelon rises to the east of Great Slave Lake, in the "land of little sticks", where the Chipewyan people once lived and hunted. It flows north and then east, through the largest game sanctuary in North America, and finally runs out onto the barrens.

The Dubawnt River flows through a portion of the barrenlands about which little is known before the first Europeans travelled the river at the turn of the century. It courses almost straight north from the upper reaches of the boreal forest, to join the Thelon at Beverly Lake. Together they head east toward Baker Lake.

The Kazan River follows the progression of the barrenlands - from the northern edge of the spruce forest to the tundra. It then winds its way north, along a course that exceeds a thousand kilometres. Eventually, its waters, now joined to the swelling current of the Thelon and Dubawnt at Baker Lake, find their way to Hudson Bay at Chesterfield Inlet, a point half way up the western coast of the Bay.

In the last two decades, small numbers of adventurous canoeists, not more than a few hundred people in all, have ventured down these rivers. Until recently, only a handful of non-Natives - explorers, fur-traders, and missionaries - had ever seen this part of the world.

Our four canoes spilled out of the river's current onto an undistinguished arctic lake. The water ahead stretched nearly to the horizon. In the distance, a thin black strip of featureless shoreline divided the lake from an immense open sky. The barely-discernible land far across the water seemed beyond our reach, almost a mirage. It was impossible to tell where the current of the Kazan would once again pick up our canoes, and carry us away from this flat expanse of lake. We paddled steadily toward the horizon, each of us alive with the anticipation of discovery.

For a week now, the Kazan River valley had defined our lives. For us the valley was an emerging land, full of mystery and discovery. We knew that the Laurentide glacier, the last remnant of the ice age on continental North America, had retreated 7000 years ago. It had left behind, in the heart of the mainland arctic barrenlands, this valley through which we were now travelling. Where there had been nothing but ice, now there was soil, vegetation, animals, and man. The layers of life had accumulated over the millenia.

As our tiny fleet of canoes edged silently out into the lake's calm, we continued to eye the thin black line that was the distant shore. A tiny obelisk appeared, no more than a slight aberration on the flat horizon. We paddled toward it, not knowing what lay ahead.

Our journey through an emerging land took our imaginations back in time, back to the era when the great glaciers of the last ice age were retreating, forming the hills and valleys all around us. We could almost see the gradual progression of life, first the establishment of tiny plants on what had been a desert-like landscape, then the reoccupation by birds and mammals that had retreated from the advancing ice. Today, those living things seem an inseparable part of the river valley. Our journey was intended to discover the processes which helped turn what was once a stony desert into a verdant wilderness.

Man, the hunter, came last, and left signs of his occupation which remain today. When the Native people lived here, their lives depended upon the other layers of life in the ecosystem. Today, Inuit live in communities and only return occasionally, on hunting forays, into the land of their ancestors. The Kazan valley now is largely an unpeopled wilderness.

On the final stretch of the river, weeks later, the canoes bounced down crests of white water. Standing waves up to a metre high became routine. When it was not contorted into a furrowed field, the water swept along at incredible speed, up to ten knots. Rocks threatened beneath the slick, smooth surface. Yet our paddles could not touch bottom. The boulder-strewn banks rushed past in a blur on either side, as the canoes hurtled down hill toward the river's mouth.

From the lead canoe, I could hear the canoes behind me slapping on the water, crashing on the wave tops. Spray swatted us. Happy paddlers let out shrieks of joy, confidence and celebration.

After 500 kilometres of paddling our canoes on the Kazan, this was the grand finale. It was a rush, a high. We had come to trust the river as we did each other, and we were keenly aware of the river's power. We had accepted the river as friend, almost as guardian. We felt we knew its ways.

Within the day, all sixteen of our canoes shot down the final 40 kilometres of the Kazan - a stretch of river with strength and character and determination that matched the new-found confidence in every one of the young canoeists who travelled it. There was no going back, no resisting. Nothing we could do would alter the river's power. As if in some kind of ritual - a coming of age - our canoes, and we ourselves, passed through these churning, excited waters to reach the inevitable end. Suddenly, the trip down the Kazan and the exhilaration were over. We had reached Baker Lake.

Later that day, we stood in a circle, near the mouth of the Kazan. We shared our thoughts and feelings, and here on the shore of Baker Lake, thirty-two members of the Canadian Arctic Expedition came to the realization that we had reached our objective. Our scientific survey of the Kazan River valley was complete. Our circle stood for the nucleus of a new knowledge. Through our collective looking-glass, we had examined the Kazan valley with the eyes of archaeologist, botanist, geologist, ornithologist, ecologist, and palynologist.

Our aim had been multidisciplinary - we wanted to see and describe this river valley as a whole. We intended that everyone on the expedition - and those learning of it afterwards - would see this remote wilderness as an ecosystem where every element of the system depended upon, and had an impact upon, every other part of that system. Certainly, we found that was true. And an understanding of that complex wholeness had spread among us. Amateurs, nearly every one, we had nonetheless accomplished a comprehensive, far-reaching, and above all multidisciplinary, scientific survey of the lower Kazan River valley. That study followed the very flow of the region's development: the soil, the vegetation, the animals, the human habitation.

Equally profound was our shared understanding and respect for the Kazan. The river had cast its spell. Each one of us knew we would carry in our hearts a special feeling toward that river for the rest of our lives. Some spoke of challenges they had met, others of the wildlife they had encountered, others of the new friendships cemented on the journey, and others of the land through which they had travelled. For everyone it was a time of connection to the river, a river they all hoped would somehow be preserved for time immemorial.

Six weeks and five hundred kilometres earlier, we had stood beside Angikuni Lake, uncertain what lay ahead. The river and its land held out a challenge. We sought to follow in the footsteps of the Kazan's few explorers. We sought to unravel some of the mystery of the barrenlands, to answer questions that remained in the minds of the few scientists who had examined this valley.

Those of us responsible for leading, (a team of eight scientists and experienced wilderness paddlers) looked at the task ahead with both eager anticipation for the value of our mission and with some anxiety for the international collection of young people in our group. Were they ready for the challenge, physically, spiritually, and intellectually? For those young people, the barrenlands mystique was daunting. It was, at the outset, an unknown for them. Only during the river journey did the land merge with their imagination and become a part of their reality.

Now, our voyage of discovery was all but over. We had travelled through a land that few had studied, a land undisturbed in its evolution to wildness as beautiful and unharnessed as any on the planet.

Our first close up sight of the Kazan valley was a lush expanse of arctic plants, flowers, lichens, tiny trees the size of shrubs. The once bare, exposed rock of the valley is now almost completely vegetated. The period since the glaciers retreated has been one of renewal and returning life, as well as sorting and fine-tuning of the ecosystem's elements.

Along one part of the river there are groves of trees, diminutive and often misshapen. They are mature, and part of a complicated and dynamic fabric of living things. Each thread of the fabric exists as an individual unit, but each is inextricably bound to the others, to form the whole. Countless insects have come to live in and on the trees, and many birds rely on these for at least part of their food. The trees themselves are important to songbirds as perches from which to declare their territories, and in providing seeds for them to eat. Some birds use the shelter provided by the dense and tangled growth at the base of the trees to protect their nests, and small mammals may also find shelter there. The birds and small mammals themselves constitute much of the food of larger, predatory animals such as foxes, wolves and raptors. The living elements of the region are interdependent.

The Kazan valley is a land that varies, in contour, in structure, in colour, and in mood. It is the barrenlands, so-called because it is a world barren of the great spruce forests that grow to the south. It is a land where the view is unobstructed, even by trees. A rise of fifty metres affords a panorama to rival many a mountain top, the land laid out below like a tapestry woven of natural hues, of greens and browns and greys and blues.

No trees meant no shade from the sweltering sun which, at times early in the trip, pushed the mercury up over 30°C. No trees meant no shelter from the gale-force winds and driving rain that accompanied weather more fitting to the image of an arctic adventure. No trees meant limited wood to build fires against the temperatures that, on occasion, put a layer of ice on the water left in our pots overnight to prevent them blowing away in the wind.

Our canoes carried us down swift currents, over great expanses of lake, through wind-tossed waves, down the Kazan River between banks sometimes high and rocky, other times low and green. But canoeing was a means, not an end. We were there to explore the river valley. The canoe was our vehicle for this journey through an emerging land.

High up the river, early in our trip, the four canoes of one group moved slowly across that unnamed lake, a body of water big enough to disguise any hint of current as the river flowed through, but small enough to have escaped the toponymist's pen. The winds that day were gentle, and paddling was easy, a relaxing rhythm. The far shore of the lake greyed into an obscure band between the water and the sky, broken only by the tiny obelisk that drew us on. From the next canoe, Jane, an ecologist from England serving now as one of the expedition's scientists, briefed us on our team's work . . .

We should be looking for a small lake off the river soon, a place suitable for taking a pollen sample in the mud lying on the bottom. If we encounter another of those isolated stands of spruce trees beside the stretch of river ahead, we ought to take some cores of the trees' heartwood. And Andrew, the expedition's archaeologist, had assigned our team another area of survey starting at the end of this lake.

As Jane talked, someone spotted two bright white tundra swans on the water ahead. We paddled quietly, hoping to see their young, and add to our sightings of nesting birds.

Every day we were mindful of the layers we studied in our examination of the valley: the glacial history, the vegetation, the animal life, and the archaeological record of human habitation. The study of one led naturally to the next, as we considered the emergence of the valley we now saw. Each layer of life depended upon its predecessors: the vegetation upon the soil, the animals upon the vegetation, the humans upon the animals.

The outlet from this peaceful lake was not yet visible on the grey horizon, featureless but for the single obelisk against the sky. We paddled on towards it, unsure why, but drawn in that direction.

Hours passed. Our paddling continued, gently urging the four lonely canoes across the lake. The far shore was clearer now. We could make out the boulders on its bank and, rising above the rest, a cairn of rocks left by some traveller long before us. It may have been a Chipewyan or Inuit hunter, marking the lake's outlet to the river. For us, that inuksuk stood as a reminder that we were visitors in a land that, at least spiritually, others before us had understood much better than we ever would. They had lived here. Their lives depended on the resources of the land. But when those resources failed them, that life ended. Today, no one lives permanently in the valley of the Kazan. It is wilderness.

The soils, the plants, the trees, the birds and mammals, and the signs of former human occupation - all of them received our attention as we travelled slowly downriver. The canoe offered an ideal perspective for such studies. No other vehicle permits the barrenlands traveller to be so close to the land. The cadence of paddling suits the wilderness rhythm. From a canoe, we viewed our environment from within, on its own level, not from high above, detached. From a canoe, it was natural to look on the land as an earlier traveller would have done, picking out the campsites and landing places with a canoeist's eye. From a canoe, it was easy to disembark, almost anywhere, to examine the land more closely. We paddled five hundred kilometres, and we walked over three hundred kilometres of the valley to either side, as we examined every layer of the emerging land.

This expedition was a marriage of science and canoeing. Each was enhanced by the other. The science forced the canoeist off the river regularly, into the hills and the willow thickets, among the rock piles, and onto the open tundra. The studies lent purpose to our inland hikes, which served up time and again the treats of the barrenlands that only those who pause from their canoeing are privileged to know. On the other hand, the canoe forced the scientist to deal with the river valley as a total system. Botany, zoology, ornithology, palynology, and archaeology all became essential parts of the puzzle - none seemed complete without the others. Travelling by canoe allowed the scientific survey to examine, in an uninterrupted manner, a large segment of this wilderness river valley. The canoe served as a vehicle of scientific enquiry.

The story of the Canadian Arctic Expedition is one of international friendship and understanding. It is one of coming to terms with true wilderness and of preserving heritage. It is one of adventure. It is one, at times, of danger, and at others of peace and serenity beyond the imagination of those who have never left the bonds of so-called civilization. But first and foremost, this is a story of scientific exploration.

The canoe has a proud heritage in Canada as a vehicle of exploration. Long before Europeans penetrated the interior, Native peoples travelled widely through the land using early forms of the canoe. Adopted by the visitors, it carried explorers to most regions of Canada. When the fur-trader Alexander Mackenzie became the first non-native to follow a river down to the Arctic Ocean in 1789, he did so in company with several canoes paddled by his Indian guides and their families. His own canoe was paddled by four Canadians: Francois Barrieau, Charles Ducette, Joseph Landry, and Pierre de Lorme. In the years that followed, the great map-maker David Thompson travelled many of Canada's waterways. In all, his journeys amounted to over 125,000 kilometres, much of it accomplished by canoe. The litany of canoe explorations goes on through the decades.

The era which saw the canoe regularly used for real scientific exploration in Canada is long gone. We thought often of that past as we paddled. But for us, on the Kazan, what mattered most was our sole predecessor in that role: Joseph Burr Tyrrell who descended the Kazan in 1894.

The Kazan that Tyrrell "discovered" was a wilderness beyond the limits of mapped Canada, a region known only to Chipewyan and Inuit hunters. Among Europeans, only Samuel Hearne, a fur trader, and Father Alphonse Gasté, a priest, had travelled part of the route before. Hearne made a remarkable journey by foot through the barrenlands in 1770. He crossed the Kazan and encountered a band of Chipewyan hunters "who had been some time employed in spearing deer [caribou] in their canoes". He makes no mention of Inuit there. A century later, in 1868, the Catholic priest Gasté entered the valley from the south travelling with a group of Chipewyan hunters who, he said, were anxious "to see the Eskimos again and to trade with them". In the intervening period, Hudson's Bay Company traders recorded reports of meetings between the Chipewyan, who traded at the Churchill post, and the Inuit along the Kazan.

Tyrrell, however, was the first to examine the Kazan with a scientific eye. Leaving from an outpost in northern Manitoba, Tyrrell's party travelled over the height of land to Kasba Lake, the headwaters of the Kazan. As he descended the river, he mapped it for the first time, documented its geology and natural history, and carefully recorded all his observations. Reading his account, there is one notable difference between the river he saw and the river today.

Tyrrell estimated the human population of the river valley to be roughly one thousand. He describes the camps, each comprised of a few families, with their caribou-skin tents and caribou meat laid out to dry in the sun. He writes of Ahyout, Hikuatuak, Kakkuk, Unguluk and Pasamut, and gives us a glimpse of life as it once was beside the river.

"Ten miles below the islands is a place called by the Eskimos Palelluah," wrote Tyrrell, describing what is now referred to as Padlerjuaq, roughly meaning the place with the big willows, "where the river is deep and narrow, and the caribou, in their migrations, regularly swim across the stream. It is probable that this is where Samuel Hearne crossed the Kazan River above Yathkyed Lake in 1770, which he describes as a celebrated deer crossing place."

Tyrrell watched Inuit hunters in their kayaks spear caribou as they swam across the river at Padlerjuaq. A century before, as the first European to penetrate the barrenlands, Hearne had encountered Chipewyan hunters at the same location, waiting to intercept the caribou on their migratory march. As we approached Padlerjuaq, was it not natural that we would ponder what we would find? What archaeological evidence of former occupation? Inuit? Chipewyan? And what signs of use by animals now, a hundred years after Tyrrell?

Equipped with an arts degree from the University of Toronto, Tyrrell was hired by the Geological Survey of Canada to do scientific exploration. He was a generalist, doing fieldwork as geologist, naturalist, and topographer, not to mention the skills demanded of him simply to survive in the arctic. His was a time of transition. The end of a century of arctic reconnaissance, of plotting the basic map of Canada's north, led to the beginning of a new era. Tyrrell belonged in both, for he explored uncharted territory, but did so as he collected data for a multidisciplinary range of scientific studies. The emphasis in exploration was changing from territorial conquest to scientific enquiry as a young nation sought to comprehend its land.

Travelling by canoe, Tyrrell examined the river valley and recorded its geological formations, its soils and plants, its natural history, and its human population. The parallel with our own approach in 1988 is striking, and that observation was not lost on the participants of the Canadian Arctic Expedition.

A few days after crossing that first lake, guided by that lonely inuksuk, we were camped at one of the wildest and most awe-inspiring places on the Kazan. It is often called the Three Cascades. In the space of three kilometres, the river drops nearly fifty metres, in a series of ledges and cascades whose beauty is unsurpassed anywhere in the barrenlands.

We stood beside the swirling water, our toes wet from the gentle back-eddy lapping up on the flat granite shelf we had chosen for our kitchen. The noise of tumbling water enveloped us. Herring gulls floated through the mist rising from the cascade. On the far shore, a peregrine falcon hunted over the tundra in ever widening arcs. On the sloping bank behind our camp, subtle shades of green were dotted with brightly coloured wildflowers at the height of their short summer bloom.

Later that night, at the darkest moment, the northern sky glowed orange and pink and the colours streaked in the rushing water. This is a beautiful place, a spot where the Kazan rises above mere scientific examination to achieve another plane of human appreciation, for its simple, aesthetic magic. Even

Tyrrell, whose writing rarely reveals the slightest emotional reaction, was moved by this place.

> *"The whole landscape, seen in the early morning light, presents such a picture of wild but quiet beauty, as I have seldom had the good fortune to enjoy."*

Before this journey into an emerging land began, I had seen it as an opportunity to give these 24 young people from around the world a new appreciation for the value of wilderness. The greatest ally one could have had in that effort was the Kazan - a valley almost magical in its influence over people.

Everyone standing at the river mouth had a deeply held belief that the Kazan River was somehow, indefinably, special. After six weeks trying to document the valley, using a range of scientific tools, every participant was left with a profound respect for the essential mystery that lies at the heart of this great wilderness. The Kazan touched each of us. As the young people talked, I heard conviction that this wilderness must be preserved. I heard pleas that the Kazan be acknowledged as a Canadian Heritage River. I heard longing for open spaces, clear water, clean air and unfettered existence. For 24 young people from around the world, the Kazan was imposing its own set of values.

For all, it was a journey of discovery, crossing new frontiers, examining the environment on every level. It was a journey rich in heritage, along a river where Native people have travelled and hunted for centuries. It was a journey through wilderness, through a land once covered in ice, but now resplendent with life. It was a journey into an emerging land.

David F. Pelly - Canada
Expedition Leader

Further Reading

Hearne, S. 1958. *A Journey to the Northern Ocean.* MacMillan and Co., Toronto.

Mallet, T. 1930. *Glimpses of the Barren Lands.* Revillon Frères, New York.

Tulurialik, R. and Pelly, D. 1986. *Qikaaluktut - Images of Inuit Life.* Oxford University Press, Toronto.

Tyrrell, J.B. 1898. *Report of the Doobaunt, Kazan and Ferguson Rivers and the Northwest Coast of Hudson Bay.* Annual Report of the Geological Survey. Ottawa.

"The cloud cover has cleared a little and we are having our first glimpse of the arctic tundra -- it looks serene and somehow waiting for us to arrive. The yellow sun - still quite high in the sky - is glinting off the hundreds and thousands of lakes dotted about the land like puddles of water on a concrete pane. Jez has just pointed out the ice covering the larger lakes - ahhhhh! watch out ice - here we come. Snow still dots the land in places - and the tundra looks a basic brown from this far up."

Sonia Mellor - Australia
Day One - July 1 (from plane)

"As we clambered out of the float plane onto the slippery rocks of the lake shore, and looked out over the land, all you think you see is endless flat land, sky, river, and lakes."

Hilli Woodward - England

THE MAKING OF THE LANDSCAPE

The Kazan is first and foremost an arctic river. Arctic! The very word conjures up images of bleak desolation. One can almost feel the biting cold driven home by the wind, and see the white and featureless landscapes, devoid of life or comfort. Visions of human despair and tragedy rise up from the swirling snow like wraiths - Hudson, set adrift in the frigid waters of the bay that now bears his name, and Franklin, vainly struggling southward to escape the hunger and cold. Beyond these images there is a deep sense of timelessness. It is as though not even nature, much less human effort, can alter the frozen mass of rock, ice and snow. The arctic lands seem as foreign as the remotest planet.

Viewed from the sprawling cities, ordered farmlands and dense forests of the temperate regions, the landscapes of the Kazan might be imagined as featureless expanses of snow and ice. However, one walk across a northern meadow on a summer day will alter these preconceptions forever. Life is everywhere. You see it in a diversity of green plants with startlingly colourful flowers. You tread upon a living carpet of moss. With each breath you can smell the rich aroma of the vegetation, the aroma of life. It would be a rare thing on such a day to not hear a bird sing or see signs of mammals, large and small. To the trained eye of the geologist or geographer, there is abundant evidence that the land has experienced dramatic environmental change, and is indeed changing today. It is richly endowed with unique geological formations, striking landforms and fascinating vegetation. Tales of tropical warmth and icy coldness are written in the outcrops of rock that border the river. Both the chilling darkness of the arctic winter and the bright vibrance of the northern summer are there for the adventurous to discover. The climatological, geological, and botanical attributes that have produced the landscape continue to draw explorers and scientists northward to the Kazan.

"About three weeks into the trip the land took over. It's really noticeable in my journal. I started to draw things like sunsets, wildlife, or just the land - walking on the tundra, perceptions, and things like that. To me that was really interesting because it was the first time the land itself, the environment, touched me so much."

Sonia Mellor - Australia

THE SOUL OF THE LAND

Climate is the true soul of the arctic. The rocks and vegetation merely provide bone and flesh that remain malleable to the whims of the arctic soul. In the Kazan valley, the arctic climate can summon cold that is capable of splitting rock and wind that can strip the bark and leaves from trees. Yet, in summer the days can last for almost 24 hours and the temperatures may reach 30°C. What then, is this thing called the arctic climate?

> *"July 7 - The rain and wind struck about midnight and stayed all day. The temperature dropped to about 5°C, and the wind, at 25 to 30 knots, whipped up white horses on the lake.*
>
> *"July 13 - Long, hot, still days of temperature 27°C in the shade and 37° in the sun. However, one must share the few protected shadowy spots with thousands of mosquitoes and black flies."*
>
> *Kassie Heath - Australia*

Sunlight is the ultimate source of energy for all life on earth. Sunlight drives the winds and sets the ocean currents in motion. In the higher latitudes the power of the sun is at its lowest and has the greatest amount of seasonal variation. That is the key to the arctic climate. In coastal regions the moderating influence of warm sea currents can partially offset the low power and seasonal variation of the northern sun. For example, Stockholm lies at roughly the same latitude as the headwaters of the Kazan, yet it experiences a far less severe climate. There are no warm seas to shield the Kazan valley against the force of the northern winter. During the winter the sun remains low in the sky and the days are very short. In December and January the dawn and sunset remain separated by one or two hours of soft illumination. Even at noon the sun lies just a few degrees above the horizon and casts low shadows across the land. With the darkness comes cold. The mean temperatures in January range from -30°C in the southern Kazan to -33°C at Baker Lake. Temperatures as low as -50°C are not uncommon along the river. Not even the swift and mighty Kazan is beyond the reach of the arctic winter. By early November the river is frozen over and the ice does not begin to break up until June.

Despite the cold and the darkness, winter is a time of unworldly beauty. On clear nights the stars sparkle seemingly just out of reach, and the full moon reflecting off the snow provides brilliant illumination of the white landscape. From late summer, over the winter, until spring, the Aurora Borealis casts swaying curtains of spectral light across the night sky. Scientists suggest that these northern lights are caused by photons and electrons from the sun interacting with the magnetic field of the earth. In Inuit legend they are arsaniit, the sky people, playing a game of ball across the heavens.

> *"Suddenly someone noticed this glowing hose overhead. The whole sky was lit up by a winding curtain of shimmering light, dancing overhead - the northern lights. I sat captivated, watching as one shade turned into another."*
>
> *Paul Clements - Jersey, Channel Islands*
> *August 17*

During the summer the sun is higher and there are just a few hours of twilight between sunset and sunrise. The soft luminance of those starless nights is the essence of tranquility. The summer midnight along the Kazan is a time to sit a little apart and let the peaceful sky work its soothing magic. It often seems as though all the animals, and the wind itself, feel an urge to be still and restful at this time. In more southerly regions the long hours of sunlight would produce lethal temperatures. However, at high latitudes the angle between the earth's surface and incoming solar rays is relatively oblique and the concentration of solar energy per square metre of land is low. Thus, the summers in the Kazan valley remain cool. The mean July temperatures are 13°C in the southern portion of the Kazan and 11°C at Baker Lake. Freezing temperatures can occur at any time of the year. In contrast, the record high during July is a startlingly warm 30.6°C.

Despite its countless lakes and bogs, the precipitation is so low in much of the Arctic that it can be classified as a desert. This is because the cold polar air masses that dominate the region do not contain much moisture. The mean annual precipitation at Baker Lake is only around 234 mm per year. In contrast, the mean annual precipitation in the heart of the arid Canadian plains at Medicine Hat, Alberta is approximately 350 mm. What then supports the multitude of lakes that spread across the landscape? Again, the key to the arctic lies in the pale northern sun. Because of the low summer temperatures, the evaporation rate of surface water is very low and the land remains extremely moist despite the small inputs of precipitation. There are approximately 127 days per year of measurable precipitation in the southern Kazan valley and 100 days at Baker Lake. Most of the precipitation along the Kazan occurs during the summer when moist Pacific air masses temporarily invade the region. However, roughly 43 percent of the annual precipitation occurs as snow.

Snowfall can occur during any month of the year. Even in the midst of summer, the winter does not completely release its grip on the Kazan.

FIRE AND ICE

"Another first today - a portage over boulders and rocks. Pre-Cambrian rocks that had been held by glaciers during the ice age. Some are rounded and almost smooth, others squared and stratified leaving strange, odd depressions"

Eddy Chong - Singapore
July 14

Some of the greatest scientific treasures and mysteries of the Kazan lie hidden in the rocks and soil. Since the earliest European expeditions, mineral wealth and geologic knowledge have remained two of the strongest inducements for arctic exploration. The 1894 survey of the Kazan by J.B. Tyrrell was one of many links in the long chain of geological exploration. The Kazan continues to hold a special interest to the earth scientist. Viewed from the perspective of the geologist or geographer the Kazan presents a unique contradiction. The ground upon which the expedition travelled is both one of the oldest surfaces and one of the youngest landscapes on the planet. Two ancient powers - fire and ice - created the central factors which determined the physical landscape of the Canadian arctic - bedrock and glaciers.

The bedrock of the Kazan River region mainly consists of granite and granitic gneiss. These rocks are produced by the cooling of molten material, referred to as magma, deep beneath the crust of the earth. As the molten rock cools, mineral crystals form. Under great pressure the minerals are fused into solid rock. When these rocks are exposed at the surface they tend to cleave and crack along planes dictated by the orientation of the mineral crystals. These long fissures are referred to as joints. The joints can extend from a few metres to many kilometres. The rectangular jointing patterns of the gneisses and granites are clearly visible where bare surfaces of these rocks are exposed today.

How old are the rocks of the Kazan? The question implies an entirely different time scale. Years, centuries and millennia are brief instants that pass almost unrecorded by the rock. Even a million years is a rather short period of time to the geologist. All of human existence, from two million years ago when our primitive ancestors began to walk upright until the present day, represents a momentary flash compared to the geologic history of the Kazan. Geologists estimate that the gneisses and the granites of the Canadian Arctic were formed over 2500 million years ago. They are among the oldest exposed rocks in the world, their formation occurring long before the dawn of life. These ancient rocks comprise a large geologic structure known as the Canadian Shield. The shield forms the solid foundation of the North American continent and Greenland. However, in most areas the rocks of the shield are covered by great depths of younger deposits. The geologic history of the exposed portion of Canadian Shield, in the Kazan valley and other parts of the Arctic, is a remarkable story of mountains, oceans and wandering continents.

Between 2500 million and 600 million years ago the rocks of the Canadian Shield were exposed as several separate land masses at the surface of what was at first a desolate and lifeless world. Earthquakes and volcanic eruptions were common on the surface of that young and inhospitable earth. Even the atmosphere would have been poisonous to oxygen-breathing creatures. During the period 2000 to 1800 million years ago the land masses of the Shield had converged, and began to move from a location near the northern pole to within 30 degrees of the equator. The movement of these huge masses of rock was facilitated by the same internal energy and turmoil that generated the earthquakes and volcanic eruptions of that distant era, and still produce earthquakes and volcanic activity today. Volcanic rocks from the period 2500 million to 600 million years ago can be found in a few places along the Kazan today. Now they are the cold and scattered reminders of a distant and tumultuous time.

Due to the intense geologic activity of the young earth, the rocks of the Canadian Shield were rent by at least three major stages of mountain building before 600 million years ago. Between episodes of mountain-building, the wearing down and transport of the rock by wind, rain and ice reduced the mountains to hills. Even the toughest rocks can be broken down to cobbles, sand, silt and clay by the action of wind and water over time. Shallow seas filled the lowlands and the fine sediment produced by the weathering of the rock was deposited in them. Over time, and under the pressure of overlying material, these deposits have been transformed into sedimentary rock. Within these rocks we find some of the earliest traces of life yet uncovered. The remains of microscopic algae, seaweeds, and sponges become increasingly common in rocks from the period 1800 million years ago onward. Some of the marine rocks dating from this time are exposed today near the mouth of the Kazan south of Baker Lake. These rocks bear silent testimony to the origins of life in the warm waters of long lost seas.

By 600 million years ago the seas supported a profusion of plant and animal life. Over the course of the next 600 million years a diverse range of terrestrial plants and animals would rise to occupy the land surface. At 600 million years ago the land mass that is now the Canadian Shield was located near the equator and experienced tropical climatic conditions. However, as time progressed the land began to move northward. By 135 million years ago large portions of the shield were located north of 60° latitude. Surprisingly, subtropical conditions appear to have occurred up to 70° North latitude. In terms of modern day climate, this would be like travelling 600 km north of the Kazan River and experiencing weather similar to that of Florida.

The mountain-building forces typical of earlier times had diminished by 600 million years ago and the surface of the shield became increasingly stable. By 500 million years ago, the topography of the shield was reduced to rolling plains and low hills. Several times seas covered the gentle landscape. Sediment deposited in these oceans was gradually transformed into dolomite and limestone rock. From 225 million years ago to the present no major inundation of the shield by seas has occurred and the sedimentary rocks have gradually been eroded away re-exposing the underlying gneisses, granites and volcanic rocks. This continued erosion of the uplands has produced a land surface in the Kazan valley that is gently rolling, with no high mountains.

During the last two million years the relatively gentle forces of wind and water have been joined by a third, far more dramatic agent of landscape change - glacial ice. By two million years ago the Canadian Arctic was located very close to its modern position near the north pole. Global climatic cooling was well under way and evidence of glacial activity can be found in deposits 25 million years old. At this time forests of spruce, pine, and cedar covered much of the Canadian Arctic. The long winter of the ice ages was about to begin.

As many as 17 times over the last two million years continental glaciers have formed in the Canadian Arctic and expanded to cover almost all of Canada and large portions of the northern United States. Similar expansions of glacial ice occurred in the high latitudes and mountains of both the northern and southern hemispheres. Glacial ice can grind the toughest rock to powder and depress the surface of the earth hundreds of metres, yet the growth of a glacier begins with one of the softest and most fragile elements of nature, the snowflake. The transformation of the delicate beauty of the snowflake into the unimaginable force of glacial ice is a story in itself.

The growth of large glaciers began with climatic cooling that allowed accumulations of snow to remain unmelted through the summer. At first the persistent snow may have been limited to north facing slopes and other protected spots. Over several years the depth of snow built up in these sheltered sites. The snow was transformed by pressure and slight melting from delicate flakes into compact crystals of ice. With time and more accumulation the crystals merged to form solid masses of ice. By this time the area of the summer snow masses had increased and individual banks began to merge and cover the land. The white surface of the snow and ice reflected much of the sun's heat and the cooling of the land increased. As decades and centuries passed, the accumulation of snow continued to create ice. Eventually, the weight of the overlying ice and snow was so great that the ice started to flow under pressure and the slow, but irresistible, advance of the ice began.

The Kazan valley probably lies close to a major centre of ice growth. During the peak of glacial expansion the ice may have been 4000 metres thick over the Kazan River and extended 2000 kilometres to the south. The ice sheet was not static. Ice flowed slowly under the pressure of gravity from interior regions towards the edges. This great mass of moving ice is called the Laurentide Ice Sheet [(1)].

The last retreat of glacial ice began with melting along the margins of the Laurentide Ice Sheet some 18,000 to 14,000 years ago. Century by century, the summers became warmer. Along the edges of the glacier the rate of melting increased. Great cascades of water flowed outward from the ice. Huge lakes were dammed against the flanks of the retreating ice and rivers were flooded with the draining waters. In places, the ice calved off the glacier forming huge bergs that drifted across these frigid lakes. Enough of the earth's water was locked in the glacial ice that, as it melted, the chemical composition of the sea was affected. More dramatically, the sea levels of the world rose.

As melting increased, the flow of the ice from the frozen centre of the continent was insufficient to replace the loss at the glacial margins. Year by year the edges retreated and the great Laurentide Ice Sheet shrank back to the place of its birth in the Canadian Arctic. The end of the ice ages came with the final melting of glacial ice in the Kazan region 8000 to 7000 years ago. The Kazan lies in one of the last areas in North America to be free of the Laurentide Ice Sheet[(2)].

The processes of glacial ice growth and decay have left unmistakable imprints on the landscape of the Kazan valley. Uplands have been stripped clean of soil in many places to expose the underlying bedrock. Not even the tough granites and gneisses could emerge from the grasp of the ice unscathed. The joints of the bedrock are riven by long grooves cut by rock debris held in the ice at the base of the glacier. Jumbled masses of boulders, gravel, sand and fine sediment, called glacial tills, fill many depressions. These are deposits of material that were ground up, mixed and carried by the ice. Large boulders, known as glacial erratics, have been plucked up by the ice and set down again hundreds of kilometres from their point of origin. Long sinuous ridges of sand and gravel, called eskers, run for kilometres across the landscape and mark the path of streams of melt-water that once flowed beneath the surface of the glacial ice. The sands of these eskers preserved some of the oldest archaeological finds along the river.

The retreat of the ice has also left its mark on the Kazan Valley. In places, the huge volumes of melt water carved channels that meander across the land like river valleys, but contain little or no flow today. As the ice wasted northward, the melt water was dammed against the icy flanks and filled much of the Kazan region with the temporary Glacial Lake Kazan. The shorelines of this ancient body of water can still be traced high above the modern course of the river. The huge cover of ice applied so much weight that the land surface was depressed under it.

As the ice retreated, sea water entered the Hudson Bay region. A large body of marine water called the Tyrrell Sea covered portions of Quebec, Ontario, Manitoba and the Northwest Territories as well as modern Hudson Bay. Waters from the Tyrrell Sea inundated Baker Lake and the lower 150 to 200 kilometres of the Kazan River. As time progressed, the land has rebounded from the depression caused by the glacial cover. However, the rebound of the land is a slow process, continuing today in the Hudson Bay region. Sand and gravel deltas formed where the Kazan enters Baker Lake when the surface of the land was much lower than today. Traces of these deposits may be found above the modern level of the lower Kazan.

The Kazan River, with its complement of waterfalls, rapids, slow water, and lakes, holds a compelling attraction to canoeists. It offers challenging runs through cold, fast rapids and then moments later provides a restful pause on flat water. The wild and diverse nature of the river is a direct consequence of the recent release of the land from the glacial ice. Phenomena such as large lakes and waterfalls are anomalies in river systems. They do not occur on rivers that have had long periods of time to develop. Over time, rivers cut down and lower the outlet of lakes and thus drain them. Similarly, erosion and incision by rivers reduces waterfalls and rapids to smooth water. The variability of conditions along the Kazan directly reflects the relative immaturity of the river. Ultimately, the very nature of the Kazan is a product of the glaciers.

Although the Laurentide Ice Sheet has left the Kazan region, the cold climate of the Arctic has accorded to ice a continuing role in shaping the landscape. During the summer, water enters cracks and depressions in rock. Then in winter, the freezing water expands with irresistible force. The expansion of the freezing water produces frost shattering of the rock. Sharp, jagged debris caused by frost shattering occurs throughout the region.

Although the winter snow cover and lake ice melts each summer, only the top metre or so of the soil surface thaws. The underlying ground remains frozen to depths of several hundred metres. This condition of continuously frozen ground, called permafrost, is ubiquitous in the Kazan region. The frozen ground forms an impermeable layer that holds water at the surface of the soil. Large areas of the landscape remain wet and at times difficult to traverse. Ice causes cracking of the ground surface and the movement and sorting of rocks and soil particles. This often produces striking geometric patterns on the surface. Nets and stripes of neatly sorted stones can develop on dry sites. To the untrained eye, these stone nets can be mistaken for ancient hearths or tent rings. In moist areas, polygons of raised organic soils, surrounded by wet depressions, are formed frequently. These polygons can produce a network of water-filled depressions that are over one metre deep. Ice remains an important part of the Kazan.

"We are sitting diagonally across from the Tyrrell cliffs. Tyrrell wrote in 1893 "precipitous naked cliffs of red till overlooked the water on the outside of the bend". The cliffs probably represent a marine strandline of the Tyrrell Sea (early Hudson Bay). The red "till" (sand and gravel deposited from glacial ice) typical of this region is salty, due to clay minerals, which attracts caribou. The terrain across from these cliffs is very rocky, but seems to be up on a pedestal."

Ashley Wooten - United States
July 29

THE LIVING BORDER

"I will never forget my feelings of awe, stepping onto the barrenlands for the first time . . . Instantly I was struck by how unbarren the land seems to be - how beautifully green."

Simon Cremer - England

"Looking closely at the tundra you see a mass of different colours - browns, reds, yellows, and greens. Plant species are numerous; birch, willow, black and white spruce, mosses, grasses, and lichens."

Hilli Woodward - England

In summer the land of the Kazan is green, startlingly green with vegetation cover that often stretches unbroken to the horizon. Given the cold climate, the frequent lack of soil and the disturbances caused by permafrost, it seems remarkable that any plants can persist along the Kazan. In fact, the region supports a rich complement of plants and bisects one of the most striking large-scale vegetation boundaries on the earth - the living border between the forest and the arctic tundra.

> *"Shade. Where is the shade? I hadn't dreamed that when so far north I'd be wanting shade. It was the first time I'd really missed the leafy canopies of large shady trees. To date their absence had all been part of my new surroundings. And it made me think of just how contrasting this arctic environment was, where just below the surface, maybe 15 cm down, the ground was permanently frozen; where the small turnover of organic matter provided little nourishment for growth; where in the winter, ice crystals whipped around the tree trunks, gnawing at their thin bark. These trees have witnessed every mood of every season of the arctic environment, unable to migrate or hibernate or change their conditions as animals are able."*
>
> *Kassie Heath - Australia*

Near the headwaters, at the extreme southern end of the Kazan valley, the vegetation is dominated by boreal forest. This is the largest forest in North America. For over 8000 kilometres, the boreal forest stretches uninterrupted from the Atlantic coast of Labrador to the Pacific shores of distant Alaska. Vast tracts of the land remain without highway or railroad line. Dark conifers, covered with black moss, surround countless bogs, while yellow flashes of autumn colour mark large patches of trembling aspen and birch. At times the quiet of the forest is torn by the flames of huge fires. Thousands of square kilometres may be consumed by the flames of an individual burn. The most common trees in the forest are the needle-leafed species: black spruce (*Picea mariana*), white spruce (*Picea glauca*), tamarack (*Larix laricina*), and jack pine (*Pinus banksiana*). Broad-leafed trees, including paper birch (*Betula papyrifera*), aspen (*Populus tremuloides*) and balsam poplar (*Populus balsamifera*) are also encountered.

The boundary between the great forest and the treeless tundra does not occur as a sharp, clean line. Rather, there is a gradual thinning of the forest and a diminution of individual trees as one proceeds down the Kazan. Only three tree species were recorded by the expedition: black spruce, white spruce, and very occasionally, tamarack. The trees are extremely scattered and very small. The maximum height never exceeds a few metres, although many of the trees are well over a hundred years old. Farther north, not even these scattered dwarfs are present. A distance of 200 kilometres lies between the relatively dense forest of the extreme southern portions of the Kazan and the last stands of trees in the north.

Climate controls the geographic location of the forest-tundra boundary. There are rapid decreases in summer temperatures and annual precipitation as one proceeds northward. The steep climatic gradients reflect the fact that areas south of the treeline are dominated by warm and moist Pacific air masses during the summer. Areas to the north are dominated by cold and dry arctic air throughout the year. The permafrost and the dry cold associated with the arctic air masses appear to be the major factors limiting the growth of trees in northern Canada.

The trees of the Kazan valley may live for centuries and the human observer may detect no change in the living boundary between forest and tundra. However, this boundary has indeed moved in the past as climate has changed. Important evidence of past vegetation and climatic change may be gleaned from studies of the forest-tundra boundary. Understanding past changes in vegetation and climate is crucial to interpreting the rich archaeological record from the Kazan. Northward shifts in the geographic position of the treeline may have encouraged the people of the forest edge, the Chipewyan, to advance down the Kazan in search of game. Subsequent southward movement of the treeline may have had the opposite effect, the Chipewyan shifting their hunting territory upriver.

One line of evidence used to reconstruct changes in the geographic position of the forest-tundra boundary is the analysis of fossil pollen grains. If forest vegetation is present near a site the pollen deposited in sediments of lakes and bogs will be from trees and other forest plants. Conversely, if tundra vegetation is dominant, the fossil pollen record will contain high proportions of tundra plants. Shifts in vegetation should be recorded as changes in the pollen content of the sediment. Extremely detailed studies have been conducted on fossil pollen deposited in lake sediments near the forest-tundra boundary in the Mackenzie Delta region of northwestern Canada. These studies show that the forest limit extended several tens of kilometres north of its modern position between 8000 and 5000 years ago. It is thought that this northward expansion of the forest represents a time of warmer summers than now occur in that region. Similar conclusions were suggested from fossil pollen studies of peat deposits in the southern Kazan region. However, several scientists have questioned the validity of the Kazan studies. Fossil soils of the type that only develop under forest have been found north of the forest-tundra boundary in the southern Kazan. Radiocarbon dates from these soils suggest that the forest limit might have been located over a hundred kilometres north of its modern position between 5000 and 3500 years ago. The expedition set out to begin resolving these contradictions.

Although devoid of trees, the tundra has fascinating vegetation. Growths of lichens provide splashes of black, green, yellow and orange on rock surfaces. Caribou moss (*Cladonia stellaris*), produces a yellow-green crust over large areas of the ground surface. In moist areas, mosses such as Sphagnum form soft carpets that stretch for kilometres. A dense cover of shrub birch (*Betula glandulosa*) runs over much of the central Kazan. Although only one metre high, the birches impart a lushness to landscape atypical of the usual perceptions of the arctic. Thickets of willow (*Salix*) often crowd lake shores and river banks, occasionally providing an unwelcome barrier for canoeists trying to disembark. Extensive grasslands form on well drained sites and provide sustenance for the muskox in the northern Kazan.

During early July, the mossy areas are ablaze with colour as the rich purple flowers of the lapland rhododendron (*Rhododendron lapponicum*) come into bloom. Of more subtle beauty are the downy white flowers of cotton grass (*Eriophorum*) that encircle moist bogs and fens. Over the course of the summer the blossoms of other herbs and shrubs add a wide array of colour to both moist and dry sites. A number of these plants can provide nourishment to animals and humans. Wild blueberries and cranberry (*Vaccinium*), bearberry (*Arctostaphylos*), cloudberry (*Rubus Chamaemorus*), and crowberry (*Empetrum nigrum*) are sweet and delicious fruits that may be found in the tundra of the Kazan region. Many other plants including fireweed (*Epilobium angustifolium*), mountain sorrel (*Oxyria digyna*), willow leaves and wild sage (*Artemisia frigida*) can be eaten as pot herbs, used as spices or brewed into teas.

How does this multitude of green life persist, and indeed thrive, in this land of wild and inhospitable climate? The richness of the tundra vegetation reflects a number of important traits that allow plants to exist under arctic conditions. Few arctic plants complete their life cycles from germination to death in only one summer. Most are biennials and perennials that persist for two or more years. Annual plants require sufficient energy and nutrients to germinate, mature and reproduce in one year. These conditions are not easily met in the arctic. To take advantage of the short summer, plants must begin growth very quickly in the spring. In the Kazan where snow cover is relatively plentiful, a number of species of shrubs and herbs retain green and viable leaves over the winter. The snow provides protection for the leaves and growth can begin as soon as the snow melts. Plants often grow low to the ground and many herbs grow in dense cushions to minimize physical damage and desiccation by wind in the winter and heat loss during the summer. Flowers and leaves may be darkly coloured to minimize the reflection of solar energy and maximize warming by the sun. In cases of the arctic poppies (Papaver) and avens (Dryas) the growth of flower stems is radially asymmetrical so that the flower is kept pointing into the sun. It has been suggested that these flowers act as tiny solar collectors keeping the reproductive organs warm.

UNDERSTANDING THE LAND

With the rich but untapped treasury of geological and botanical knowledge available from the Kazan it was difficult to know where to focus the efforts of the Canadian Arctic Expedition. Eventually, the botanical studies came to centre on the examination of the past and present vegetation of the forest-tundra boundary. Despite scientific interest in the ecological and climatological aspects of the Canadian treeline, there is virtually no information on the details of modern distribution of trees in the middle Kazan or on the history of treeline movement in the region. The expedition members made scientific observations and collected samples to answer three important questions.

First, what is the actual geographic limit of tree species in the Kazan valley and what species of trees and tundra plants are present in the region?

Second, are the trees increasing or decreasing in abundance? Climatologists believe that the Arctic may be one of the areas of the world most sensitive to changes in temperature caused by the greenhouse effect. If climate is warming, we expect to see an increase in the abundance of trees in the arctic.

Third, is there evidence in the fossil pollen record to suggest that the limit of trees has been located farther to the north or south at any time since deglaciation? Any past changes in climate and associated movement of the geographic position of the forest-tundra boundary undoubtedly had a significant impact on native peoples in the Kazan region. Documenting such changes is of central importance to interpreting the archaeological record of the Kazan.

Basic information on the distribution of trees and tundra plant species along the Kazan from Angikuni Lake to Baker Lake was obtained by recording visual observations and collecting herbarium samples of plants. In places, expedition members took to the high ground with a telescope and maps to chart out the living panorama of trees and tundra. In other areas, rapid mental notes were made while coping with the more pressing demands of running through white water. When time and the abundance of specimens allowed it, collections of plants were made and carefully stored. More than fifty different species of plants were found by the expedition. Three species of trees were observed - white spruce, black spruce and tamarack. Tamarack, however, was represented by only two individual trees. Many of the spruces occurred as low dense shrubs rather than upright trees. Trees higher than three metres were very uncommon. The density of trees ranged from extremely compact stands in favourable sites to a few scattered individuals in most areas. The northernmost stand of trees occurred on the southwestern side of Yathkyed Lake. Both white and black spruce were found at this site. Beyond this last stand, the tundra was devoid of even stunted trees.

In order to gauge if the trees along the Kazan were increasing or decreasing in abundance, detailed studies were made at six stands located between Angikuni Lake and Yathkyed Lake. The work was not always comfortable or easy. On warm days black flies and mosquitoes converged en masse to feast on the expedition members. The slow and careful measurements required by the studies were not always easy in the face of these insect pests. On cool days, the biting wind kept the insects at bay, but cold hands and flapping notebooks made work even slower. It was often closer to midnight than supper-time when the expedition members returned to camp for dinner after working on the tree stands. Experiencing the changeable and often harsh conditions gave a deep appreciation for the tenacity of these hardy northern trees.

The efforts of the expedition members resulted in large amounts of scientific data. Information was collected on stand density and the presence or absence of seedlings. Tree ring cores or disks were recovered from 210 trees at the six sites. Each summer trees grow radially at the trunk. The cells produced during this annual phase of growth result in distinct concentric rings. Counting of these rings from the centre of the tree trunk to the bark provides the age of the tree. If the trees along the Kazan have been reproducing annually, we would expect some of the trees to be only a few years old and some to be over two hundred years old. If trees have not been reproducing in recent time, we would expect to find only very old trees. Obtaining a tree ring core does not significantly injure a tree. Taking disks is lethal, so only trees that were simply too small to core were disked.

The only new seedlings found occurred at the northernmost site near Yathkyed Lake. It is possible that despite its northern location, some aspects of microclimate or local soil conditions may make this area more favourable for seed regeneration than sites to the south. It was noted that many of the black and white spruce were established not from seeds but from the growth of branches of adjacent trees that came in contact with the ground and established their own root systems. Preliminary counts suggest that trees from the Kazan valley with trunks only 5-10 cm in diameter can be over a hundred years old.

"I jotted down the data as someone called out the figures - spruce tree species, height, diameter, presence of male and female cones, mature or immature tree, and the shape that the tree had grown. We had become very efficient at recognizing these particulars, a task that had been very slow on our first stand. As the figures accumulated they created a picture of a typical spruce tree - 150 cm tall, 10 cm diameter, sprouted as a juvenile layer from the roots of a mature tree. Only in the most northern grove had we discovered seedlings: sprouted from cones that almost all mature trees produce but by which means they seldom reproduce. Such were their struggles!

"As our typical spruce tree emerged from the data, each tree took on an individual character. One such tree I distinctly remember. 'Diameter 22 cm,' someone said. 'What, 22 cm?', I questioned as I pulled the foliage aside to inspect the trunk. The tree was barely a metre high and even trees twice its height had trunk diameters well below 20 cm. Under its low spiky foliage, a thick, heavily barked trunk emerged upright, then immediately bent horizontally. The old trunk appeared as a hunched cripple, clinging to the warmth of the ground. We hesitated, staring at this bent old tree. 'How old?' we puzzled. We cored it carefully, respecting its age, feeling a little guilty that we were inflicting another trauma on its history of hardships. Tentatively the core emerged . . . each fine stripe on the wood indicating another year. And then, rot. A moment of disappointment was quickly shelved. Despite its rotting core, the tree managed to sustain healthy foliage. The old, twisted tree

was dying, a slow dignified death, the silent death of aging. And science gave no answer to the secret of its age."

Kassie Heath - Australia

In addition to the examination of modern trees, the expedition contributed to the reconstruction of treeline history from fossil pollen records in two ways. First, cores of sediment were recovered from two small lakes in the southern Kazan region at Ennadai Lake, and a small lake near the north end of Angikuni Lake. The actual coring of these lakes provided both exciting adventure and backbreaking labour.

"This landscape is so different to that preconceived idea I had before coming. Although I hadn't been expecting a snow-covered waste, the words arctic and barrenlands had left a sizeable impact on my psyche. So, when I stepped off the plane at Ennadai Lake, the shock was considerable. Instead of grey rock and expanses of sparse, barely living vegetation, I found a rolling green landscape with a lake of the deepest blue-green, a sandy esker containing pink hues tipped with trees."

Simon Cremer - England
June 30

At Ennadai Lake a small party was set down by float plane, the first part of the expedition to land on the barrens and the farthest south that expedition members conducted scientific investigations. It was a glorious June 30th. Temperatures were near 25°C and a brisk breeze kept the mosquitoes at bay. The sands of a large esker system provided a brilliant pink beach for landing. Arranged along the flanks of the esker were several small lakes. The deep blue of the lakes provided a striking contrast to the pink sands.

We paddled and portaged two canoes to the lakes that we wished to core. On the banks of each lake we linked the canoes together in catamaran fashion. We then anchored our strange looking craft in the centre of the lake and proceeded to take sediment cores using a light weight drilling apparatus. At the end of the day we radioed our success to the newly established base camp at Angikuni Lake and were soon flying north to join the rest of the expedition. It was a rude shock to arrive at Angikuni Lake to find the temperature only a few degrees above zero with a roaring wind giving extra teeth to the cold.

"Mike had to come and get us out of our tents so Dr. Glen MacDonald could get his core samples. After a somewhat short portage, we got all the canoes lashed together and began our journey for the core. After many tries, Glen decided it was time to go to the next lake. After repeating the same process, we went to the next lake. Tempers were beginning to flare, and I was getting cold and tired. We finally got a decent core by Glen's standards and then he said we had better get another one to make sure."

Ashley Wooten - United States
July 1

On July 1, another small group set out in cold and fog to secure a sediment core from a small lake near the Angikuni base camp. We left in the morning on what we assumed would be a short trip to a lake located about half a kilometre away. After a portage over patterned ground and dense shrub birch, we assembled our two canoes into a catamaran and set out to obtain a sediment core. Unfortunately, the lake proved to be very shallow and possessed no sediment except sterile rock. Cold, but optimistic, we set out for another lake another half kilometre distant. Soft ground and birches that seemed to grab and hold our legs made the portage slow and difficult. Once more we put our canoes together and set out to core. Again we found the lake shallow, rocky and sterile. Noon had now passed and the cold was joined by hunger. Bits of salami and chocolate did little to ward off that empty feeling.

After a long and tiring portage we arrived at the shores of a third lake. Here we found the water deep and the sediment rich with organic matter. However, the wind was so strong that our anchors could not hold us without the assistance of paddling into the breeze. Thankfully, our efforts paid off in two good cores. As we headed back a float plane passed overhead bringing the final expedition members from Churchill. We reached camp at supper time, cold, hungry and exhausted.

The fossil pollen from these sediment cores will record changes that have occurred in the treeline vegetation over time. Age control for the cores will be provided by radiocarbon dating of organic matter in the sediment. The analysis will likely take several years to complete. The Kazan gives up its secrets only to those willing to invest hard work and patience.

The second way in which the expedition began to reconstruct the treeline history involved periodic sampling at locations removed from the river itself. Samples of lake-bottom mud were obtained from nine small lakes between Angikuni Lake and Baker Lake. Detailed observations were made on the vegetation surrounding these sites. The acquisition of these lake sediment samples was often difficult. The lakes could not possess inflow and outflow streams. Often, considerable portages were required to reach the right lakes. In some instances the first lake encountered was too shallow and did not possess suitable sediment. When a good lake was finally reached the expedition members were rewarded with nothing more than a bag of dark green mud. The pollen from the modern lake muds will be analyzed and compared to the present vegetation to determine how the tundra plant cover is reflected in pollen records.

"July 4 - We did our first lake sampling today. We had a bit of bad luck with this. First, we found that all four of our plastic cups have been broken, so the all-purpose duct tape was put to use again. Then we found that the first lake we visited was not deep enough. Only on the third lake we visited did we find it deep enough to collect the sediment samples.

"August 4 - The first lake was a real disappointment. It was too large and too shallow. I would not have believed it if I was not there myself. The next possible lake was downstream of a set of rapids, which we had to portage, and on the other side of the river. It was a hard ferry across the river just below the rapids. Fortunately, on this second attempt, the lake was deep enough and we were able to collect the required sediment samples."

Boy Mow Chau - Singapore

REFLECTIONS ON AN EMERGING LAND

"The ecosystem's fragility is appreciated when you realize you've come across only two arctic poppies over a distance of 500km."

Hilli Woodward - England

There are countless unpleasant pre-conceptions about the arctic and its "desolate" landscape. In fact, the great antiquity of the bedrock, the youthfulness of the land emerging from the grip of glacial ice, and the fascinating nature of the vegetation, all combine to create a landscape of striking visual appeal and compelling scientific interest.

The land and its vegetation are fragile. Disturbing the soil surface in permafrost regions can leave ugly sterile scars that take decades to heal. Despite its abundance, arctic vegetation clings to life very precariously. Many decades may pass before vegetation can recover from destruction and removal of the plant cover. Fires built against rocks or the movement and rearrangement of large stones can destroy natural lichen cover that has taken hundreds of years to establish and grow.

It is likely that in the years ahead increasing numbers of hardy individuals will be drawn to the Kazan valley by its remoteness and challenge. Interest in geology, glacial history and vegetation will undoubtedly continue to lure scientists to the region. Only if these future venturers continue to treat the land kindly will its remarkable beauty be preserved for later generations, so that others can seek to enjoy and understand this emerging land.

Glen MacDonald - Canada
Expedition Palaeo-ecologist

Further Reading

Bird, J.B. 1980. *The Natural Landscape of Canada: Second Edition.* John Wiley and Sons, Toronto.

Fulton, R.J. and Andrews, J.T. (editors). 1987. *The Laurentide Ice Sheet.* Special Issue of the Journal 'Geographie Physique et Quaternaire' vol. XLI. Les Presses de l'Universite de Montreal, Montreal.

Johnson, K.L. 1987. *Wildflowers of Churchill and the Hudson Bay Region.* Manitoba Museum of Man and Nature, Winnipeg.

Pruitt, W.O. 1978. *Boreal Ecology.* Studies in Biology no. 91. Edward Arnold, London.

Redfern, R. 1983. *The Making of a Continent.* Times Books, New York.

Ritchie, J.C. 1987. *Postglacial Vegetation of Canada.* Cambridge University Press, Cambridge.

Walker, M. 1984. *Harvesting the Northern Wild.* Outcrop Ltd., Yellowknife.

"I think I'm finding the 'nothing' in this country of miles and miles of nothing. The minute, the hidden, the unchangeable, the dramatic, the shy, the larger than life. The delicate beauty and intricacy of the arctic blossoms, the never-ceasing flow and noise of the river, the winds, the rains, the sweltering heat, the hills, the bogs, the rocks, the sky, the ever-present light."

Leslie Mack-Mumford - Canada
July 18

WILDLIFE INHABITS THE LAND

Shortly after the retreat of the last glacier from the Kazan River region environmental conditions moderated to the point that animal life could be sustained. Initially the land had offered very little. The ice-scoured rock lay bare. But even in that vast plain of cold rock, there were small crevices which retained airborne particles of life-giving minerals. Lichens, the vanguard of natural colonization, were able to derive sustenance from this material, and from the rock itself. A natural succession began. Wind-blown material accumulated, and the rate of the process was increased by the irregular surfaces of the lichen plant bodies. Dead lichen colonies provided more nutrients, and an incipient soil began to form. In time, mosses, grasses and herbs were able to thrive where there had been only bare rock. Later, the new pastures would be colonized by birds and mammals.

Under the cover of the winter's first snow, a small brown lemming was busily collecting grass and other fine stems which he needed to build a winter home. Now that winter had truly arrived there was no time to waste. He must gather his materials and set to work, carefully weaving hundreds of delicate stems into a strong, hollow ball about the size of the space formed when a man cups his hands together. The ball would be large enough to hold his family, huddled together for warmth, and would have a small circular entrance hole. To complete his home he would line it with fur, gleaned from wherever he could find it.

The winter nest of the lemming in the near zero degree world below the snow is his key to survival through the winter. Because of his diminutive size he cannot grow a denser winter coat; a thick coat would restrict his activities to a dangerous extent. One of his responses to the problem is to create another means of insulation - a fur-lined nest.

The project takes some time to complete, and as he roams around under the snow in search of suitable stems, he forages for food to keep up his strength. Feeding is a time-consuming business, using up a significant portion of his day. But his instincts also tell him that he must eat properly. Without frequent meals his energy would soon wane, for his small body cannot store very much reserve fat. So he scurries around, eating at frequent intervals when he comes across a suitable patch of vegetation, and collecting other vegetation for his house.

A few centimetres above his back, just above the surface of the new snow, the temperature is dropping. By mid-winter it will have plummeted to minus forty Celsius, with even colder snaps at intervals. But the snow traps air and provides a milder environment, with temperatures remaining near freezing. Below its blanket, the lemming enjoys constant cool conditions throughout the winter, safe, so long as the snow remains, from the icy blasts above.

In the first few hours after the snow falls, it undergoes changes. The layer next to the ground thaws, partly because of the warmth from the living vegetation, and partly because of the physical laws governing the behaviour of water molecules. It refreezes in a denser form below the overlying layer. Because of the change in density, it effectively shrinks, thus creating a perfect living space for the lemming, with a hard bright ceiling and a carpet of forage. As more snow accumulates above and the days shorten, his range darkens, and he must rely more on touch and smells until the sun returns and the snow thaws. His survival depends on remaining under the snow until the air temperature once again increases to the point where it does not sap more of his body warmth than his coat can insulate against.

Once his nest is complete, all the lemming has to do is eat and sleep until spring returns. He joins the huddled mass of warm bodies, comfortably enveloped by the extra warmth they provide.

Each spring, as the blanket of surface snow thins, a certain amount of light and heat filter through to the vegetation below. This cue awakens the dormant plants. As the temperature warms a little more, they resume their own activities of food production, growth and reproduction. And so the cycles repeat themselves, as they have done several thousand times in the Kazan valley since the glaciers disappeared.

It is possible that small mammals such as lemmings were among the first animals to move into the Kazan valley as the last glaciers retreated, as soon as there was sufficient vegetation available to support them. There is some evidence to suggest that the numbers of small mammal species (shrews, mice, voles, lemmings, hares, ground squirrels) in the Arctic have increased since the last glaciations. This is likely the result of an influx of animals from refuges elsewhere, which would suggest that these small animals are able to adapt quickly to new habitats. When they first arrived, there would have been no threat from carnivorous predators, which require a plentiful supply of food.

Lemmings and similar small mammals are thought to have originated in Asia, long before the most recent ice age. It is possible, then, that colonization of the Kazan valley area was actually the recolonization of a land they had inhabited previously. In any case, they moved in from the periphery of the ice sheet, from ice-free refuges where they lived during the maximum extension of the glaciers in North America.

Like the lemming, the distant ancestors of the muskox originated in Asia. They made their way across the Bering land bridge about 125,000 years ago. During the warm interglacial periods of the Pleistocene, the animals roamed the tundra along with woolly mammoths, mastodons and other large, now extinct, mammals. Muskoxen are living relics of this long ago geological era.

Each time the ice returned they found themselves forced out of their habitual area and into other lands to the south. When the ice once more receded northwards they followed close behind its trailing edge. They probably arrived in the Kazan area after the last glaciation as soon as the vegetation was abundant enough to support them. Of the modern day animals now living in the Kazan valley, muskoxen likely have witnessed more of the comings and goings of new settlers, and seen more of their successes and failures, than any other species.

A lone muskox browsing on a small willow bush in a wet area at the edge of the river was suddenly distracted by something which flashed quickly in front of him. He looked up, but saw nothing. Not an excitable beast by nature, he returned to his browse. He continued to feed on and off throughout the morning with no further distraction. Then as he lay down on the drier ground of a ridge overlooking the river, another flash caught his attention. Not long after, a patch of sky above the river filled with scores of small birds. The muskox looked on, with a mild, passive curiosity. He had been living in the area for several years, but had seen very few birds there and never a cohesive flock.

The birds were following the line of the river. Just north of the muskox they reached a suitable place, far enough away from such an unfamiliar and, so far as they could tell, potentially dangerous creature as the muskox. They dropped lower and left the shiny silver thread which had guided them north. Now they dispersed over the tundra. Some disappeared into the distant haze, others alighted on top of small bushes where they could survey more closely this new area in which they would make their summer home.

Over the next two weeks, the pairs of birds, he with his jet black cap, face and throat, his rich chestnut brown nape and his white belly, and she with her less flamboyant nape and generally more drab appearance, busied themselves in preparation for nesting and rearing their young. The two Lapland longspurs found a suitable nest site on a dry mossy hummock, with protection provided by a heather bush. There they built a small cup nest of woven grasses, and lined it with finer grass and some insulative cotton from the early-flowering willow bushes in the area. It was finished in time, and when the female laid her four eggs she could incubate them safely, concealed for almost two weeks below the bush. The clutch was fairly small, and the parents successfully reared all their young to fledging before heading south again for the winter.

The other members of the flock also had a successful year, and so the area became established as a breeding range for Lapland longspurs. Countless generations returned in later years, and most, but not all, fledged healthy broods.

Distant relatives of those first birds returned to breed in that peaceful area just inland from the shore of the Kazan River many generations later. Their nest complete, the female laid her six eggs and began to incubate. She had been sitting for three days, through two severe summer storms which had left her colder than was good for her, though not wet. It was proving to be a difficult year. On the fourth day the weather was clear. She sat in the dappled light below her willowy shelter, listening. It was not clear to her what the sound was, but she could tell something was approaching. From her low vantage point she could see nothing, and in any case she knew that she must not leave her nest. Her mate returned with some food, and then perched high above her on the topmost twig of the willow. He was listening too. As she waited, the distant noise grew louder, and then she felt the ground beneath her nest begin to vibrate gently. A few seconds later her mate called out in alarm, and she left the nest to investigate. She barely had time to fly up into the bush before a small band of wandering caribou walked by. The birds escaped to a safer place to watch. When the animals had passed they flew back to the nest, now just a few twigs trampled on the tundra.

It was doubtful if the season would be sufficiently long to begin again. The weather so far had hinted at a short, meagre period for breeding, and now that the caribou had altered their migration route slightly and had begun to move through the area, the site was no longer safe.

The caribou was unaware of the havoc she had wreaked on the longspurs. As she walked swiftly on, the sun giving her summer coat a deep brown warmth, she had only one preoccupation: to reach a suitable area in time to drop her calf. She was feeling very tired as she neared the end of her long journey, but she was urged on by the scores of other animals moving in the same direction, and with the same urgency, all around her. She had made similar journeys northward across the tundra many times. Each spring was the same. Off they would go, heading north over the frozen ground and across icy lakes, always on the move. Some of the animals had spent the winter nearby, but others were strangers. Each year it was like this: some new, some old familiar animals. All had the same inner drive. At the northern end of their journey, the groups of females would gather to calve.

The old cow had noticed a change over the years, a move to a slightly different area. The changes could not be described as significant, on their own. The cow could not know that the same pattern had been repeated for many generations. When the small, occasional changes in the final destination were seen collectively, it was clear that they formed a gradual northerly extension to the caribou migration. This coincided with the gradual warming of the climate and an improved food supply. The improvements lagged behind the retreat of the ice from the area, but as the southern extreme of the glacier moved north and left behind the newly formed landscape, so, in time, the caribou followed.

Caribou have been in evidence in tundra areas since the beginning of the Pleistocene epoch, about two million years ago, when they shared their home with woolly mammoths and huge versions of our present-day moose. It is believed that caribou evolved their northern, migratory pattern on the North American continent. They had moved across the land bridge that existed between what is now Alaska and Asia when the sea level dropped during the Pleistocene. They, too, came back to the Kazan region when the glaciers left.

Fossil evidence from Eurasia indicates that caribou have formed a vital part of human culture since earliest times, even before the emergence of modern man. They are also a part of the earliest North American cultures. There is no doubt that not only the value of the meat but also that of the skin and other parts of the caribou, so highly adapted for a northern existence, were recognized very early on in man's colonization of the tundra.

Now the cow could go no farther. She sank down and, for the first time, she rejected the demands for attention from her yearling calf. The young animal which had accompanied her throughout the long journey was disbelieving at first. It continued to ask for comfort from its mother but received repeated cold head-butts and warnings in return. Eventually it gave up, and wandered off alone among the other expectant mothers scattered on the tundra. The cow had arrived at the place where she would give birth to her calf, and now she must give her full attention to that task.

Over the next two days she did not wander very far, but spent her time feeding on the tender young herbs around her and resting, rebuilding some of her lost strength in readiness for the time very soon when she would need it again. Then she dropped her calf: a young bull. She immediately cleaned the youngster and all signs and smells of its birth, an instinctive response to the ever-present threat of predators. This act meant consuming everything from the birth except the calf itself. This area was not so obviously populated with dangerous predators as previous calving areas. But her cautious instinct was inherited from all the generations before her, and so she continued to clean up. Within minutes the calf was up, supporting himself none too confidently on wet, bending legs. The air was still but very sharp. And after a few seconds he gave up the struggle to stand and sagged into the clump of grass where he had first met with the tundra.

Throughout the day, at intervals, he rose again to his feet, and gradually the shakiness disappeared and he was able to take a few steps away from the cover of his grassy bed. It was not easy, for the tundra was not as flat as it seemed. He looked out over it to see his mother feeding and, in the distance, many more caribou adults and calves. His mother was constantly mindful of him, alert to him and everything around. She would often move closer to him, making soft grunting noises of attachment and encouragement. By the end of the day his strength had improved to the point that he was able to follow his mother around as she fed. This was a tense time for the cow, because the youngster was completely trusting of everything and very inquisitive.

It would be a full-time task for the cow to protect him from his own inexperience, a task which would probably last through the summer as they roamed over the tundra together and met up with the older animals and mature bulls who had followed them north.

It is thought that the ancestors of the wolves which today inhabit the mainland tundra persisted in a relatively mild refuge, situated in Alaska and eastern Siberia, during the maximum extension of the ice. They returned to the tundra when conditions improved. It is probable that the caribou, which are true arctic animals, were already present, so it is conceivable that the wolves simply moved in to resume their familiar hunting habits.

Today, the annual pattern of movement of the caribou appears well established, with small changes in the animals' routes occurring from time to time. Such events are often attributed to a peculiar weather pattern or a disturbance of some other kind somewhere along their migration route.

When the ice first left the region, the variation in the weather from year to year was probably much more extreme than it is now. Animals seeking out the freshly available land and settling there would probably have experienced severe short-term changes in conditions. Accordingly, occupation would occur gradually, fitfully, and over a period of many centuries. Some years would be good, the vegetation would be productive, and so too were the herbivores that fed upon it. In other years the weather might go into a decline and the plants succumb to a particularly cold season. In such years the herbivores would suffer also.

Later in the successional process, when the land was sufficiently well stocked with prey animals to support carnivores, a bad period would have an effect on their numbers too. As long as the ice was not far distant, the short term, sharp effects of its fickle behaviour would be passed on and felt directly by the developing ecosystem along its fringes. Thus colonization did not occur in a single smooth, continuous progression, but was patchy, episodic, and above all, gradual.

It can be said that the present inhabitants of the Kazan valley show advanced adaptations to their environment. These adaptations have not been acquired since the last ice retreated, but rather have evolved over the total period that the birds and mammals have been living in such extreme, wintry, northern conditions. The winter-adapted animals we see today on the tundra beside the Kazan are those which have evolved gradually and were able to survive in unglaciated refuges for the duration of the ice ages. After the last ice sheet retreated from the Kazan region 7000 years ago, these animals were able to recolonize the area. Conditions had reverted more or less to those which existed before the ice extended. There was a lag because a glacier scours the land clean, and once the temperature improved there was a period before vegetation could once again become established. So the animals have been less than 7000 years on their new land. Over this period pre-adapted creatures moved back from their refuges in Alaska, Siberia, the northeastern United States, the Canadian high arctic archipelago, and Greenland.

What else has happened here since the ice left, and how is the land maturing? We have some indications from observations of birds and mammals during our journey.

As in any ecosystem, the plant, bird and mammal inhabitants of the Kazan are dependent upon one another in countless ways. The complexities of these inter-relationships evolve slowly from an initial mix of raw ingredients. In a sense, the Kazan system was already part way along its evolutionary path when it grew up after the retreat of the glaciers, because the birds and mammals which moved in were already in existence in their present form and living in adjacent refuges. Since the glaciers retreated, the many elements have gradually come to occupy their own stable, though not static, niches. As we travelled through, the land revealed some of its secrets to us, little by little. Gradually we pieced our disparate experiences into something of a whole. Slowly, we began to understand the emergence and maturation of life in the Kazan valley.

Jane Claricoates - England
Expedition Ecologist

Further Reading

Banfield, A. W. F. 1974. *The Mammals of Canada.* University of Toronto Press.

Lopez, B. 1986. *Arctic Dreams.* Picador.

Sage, B. 1986. *The Arctic and its Wildlife.* Croom Helm.

Arctic Wildlife Sketches series. NWT Renewable Resources. Information pamphlets on various species of northern wildlife.

"The big event of today happened as we were scouting a set of rapids. The eight of us were just coming up a small rise when we came upon a herd of muskoxen grazing. I was looking down at the ground when suddenly there was a loud rumbling of hooves. The herd must have been as startled as we were, but they quickly got themselves organized and formed a line of defense in front of us. We counted three males, 16 females, and one calf. There we were, the red gore-tex army facing the furry brown army. We were only about 30 metres apart. You need not be a brilliant military strategist to deduce that our best course of action was to make a hasty retreat. As we walked back swiftly to our canoes the herd followed slowly behind. Then we came back, fully armed ... to shoot with our cameras."

Boy Mow Chau - Singapore
July 30

DISCOVERING BIRDS

I had been stalking this mystery creature for what seemed like ages. The rest of the group was busy sampling a lake bottom for pollen as part of the treeline study. I wanted to see if there were any birds around. Three female Oldsquaw skittered out of range of my binoculars as I passed their pond. Almost immediately a plaintive "quoodle!" came from the edge of the pond and continued incessantly. I scanned the sparse vegetation of the shoreline eager for a glimpse of the ducklings I was sure were hiding there. Up until now, we had been unable to assign to waterfowl any breeding evidence more significant than their presence in suitable habitat. I was getting annoyed with these ones for being so perfectly camouflaged, and turned my head from side to side in an effort to pinpoint the sound. My time was growing short; we had stopped here only to do the lake bottom sample.

I decided to walk over the ridge beyond the pond to look for other species. With luck, I would surprise the ducklings on my return past the pond. The "queedle-quoodle" stayed with me, and I no longer had any idea from which direction it came. Then a piece of tundra ten metres to my right caught my eye as it ran a metre and then stood and "quoodled" at me. My prospective ducklings were now positively identified as an adult lesser golden-plover, resplendent in full breeding plumage. Its back, spangled in gold, black and white, was cryptic against the backdrop of dry upland tundra. A white band separated its mottled back from its black face, throat and belly, giving it the striking appearance of a pharaoh in regal head-dress. Aside from its good looks, this species is also worthy of admiration for its flying ability; the lesser golden-plover is a marathon migrator.

When their arctic nesting season comes to a close, these plovers head for Labrador where they congregate and feed for the next phase of their journey. A non-stop, 4000 kilometre flight over the Atlantic Ocean brings them to Brazil, and from there they continue their trek to their Argentinian wintering grounds. When spring urges them northward again, they travel over Central America, then along the Mississippi River valley to their summer breeding grounds which extend to the most northerly islands of the high arctic. The round trip journey of the lesser golden-plover, a bird which weighs about as much as a medium-sized apple, is an amazing 25,000 km. Without even considering the phenomenon of navigation, it is difficult to understand how birds are capable of accomplishing such a feat. Migration has long fascinated and puzzled man, and numerous studies have examined the many physiological adaptations birds have made to accommodate these gruelling flights. An equally puzzling question, however, is why do birds migrate, and what initiated this behaviour?

The origin of migration is subject to nearly as much speculation and theorizing as is the origin of the human species. It is widely accepted now, however, that migration occurred before the great ice age of the Quaternary Period, two million years before present. Many of the species we saw along the Kazan follow migration routes that correlate to the path of ancient waterways present before the last ice age which have now disappeared. In past geological times, the world climate was much warmer than today, providing a suitable environment in the north for a resident bird population. As the Quaternary glaciers advanced, birds were forced to move southward to survive.

Once the land re-emerged and the climate began warming with the retreat of the glaciers, many species returned to their ancestral homes in the north. But, because the post-glacial climate continued to deteriorate and improve on a seasonal basis, the migratory lifestyle was enforced. Our lesser golden-plovers fly from the food-depleted tundra to the rich beaches of Argentina every year; there they live incognito in their drab winter plumage, like movie stars seeking some peace.

As I viewed this tiny segment of the plover's life, it seemed impossible that it could be here in the Kazan valley at all. I remembered that, somehow, the birds hatched this summer would make the voyage south unaccompanied by the adult birds. They would arrive at the traditional wintering area having never before made the journey. Most would also return to the vicinity of their hatching the next season. Perhaps some would be blown off course, others stumble on suitable new breeding areas, or out-do their fellows by forging farther north. Over vast periods of time, species may evolve, disperse, or become extinct. The component species of an area's avifauna is continually changing. Although the ancestors of the birds we saw may have been present along the Kazan for centuries, it is likely these species originated elsewhere.

The loons we saw performing their courtly minuets are members of the most primitive genus still living. They, and many of the species we found in the Kazan valley, are common throughout the north of both Eurasia and North America. Indulging their preference to travel along a coastline, they may have followed the Bering land bridge from the Old World and spread into North America's similar habitat. The origin of other species, like the lovely red-necked phalaropes, may always be a mystery. Because these birds range the ocean in winter, it is difficult to determine where their starting point may have been; the ocean provides no clue to a plausible travel route. Perhaps, like Aphrodite, they were born from the foaming crest of a wave . . .

What is it that incites so many species of birds to endure the hardships of migration, to leave their seemingly comfortable winter homes and return to the arctic barrenlands of the Kazan valley?

Because the tropics were not affected by the last glaciations, tropical ecosystems have had long periods of time in which to develop intricate and diverse food webs. If we had been paddling down the Amazon River instead of the Kazan, we would undoubtedly still be trying to identify the plethora of bird species breeding there. Conditions for life in the tropics are favourable; the climate is relatively stable from month to month, resulting in year round availability of food. Similarly, reproduction can occur almost at leisure over the course of the year with correspondingly high survival rates. This tropical paradise, however, is not free from problems. It is a crowded area, with high levels of competition for food and territory.

In contrast, the barrens provide vast quantities of available space and, in spring and summer, virtually limitless supplies of insect food. Long daylight hours lengthen the feeding time each day to assist the birds in adding the reserves of body fat that will fuel their southward flight. Although the migratory journey to the Arctic is energetically costly, once there, birds are relatively free from competition with other individuals and species.

Although living in the Arctic reduces competition to a manageable level for part of the year, there are many other factors with which birds must cope: unpredictable weather; a generally cold climate; a short summer season. All of these factors can be linked to northerly latitude, and all test the ability of a species to adapt. Of 426 bird species which nest in Canada, only 66 of these use the eastern barrenlands as a breeding ground, and a mere five, in most years, overwinter.(3) By developing ways of efficiently storing heat, these species are able to exploit the resources that remain when other species have left.

Walking along a large rock outcrop, thinking about the water pipits we had seen not long ago, I was surprised when the first person in our line abruptly stopped at a lip of rock. We quietly edged toward the spot, where the sudden halt was explained with a single, whispered word: "Ptarmigan". In a sunny, sheltered nook a metre below us, a female clucked gently while animated fluffballs, her chicks, scurried to the security of her side. The adult watched for our next move, the presence of her young checking her desire to fly off. This gave us a chance to study her in detail, looking for the clues that would reveal to which species she belonged.

There are two species of ptarmigan in the area of the Kazan, and their similarity makes it difficult to distinguish between them. We were most familiar with the russet-brown willow ptarmigan which we encountered close to the river's edge in the willow-covered places it prefers. Because of the high rocky location and the dusty brown plumage, this one, we decided, was a rock ptarmigan. The seven downy chicks, waiting for the next cue from their mother, were quite capable of feeding themselves despite being not much more than a week old.

Ptarmigan are a model of thermal efficiency. From our vantage point, we could clearly see the feathering that extended over the legs leaving only the toes exposed. Feathers provide insulation by trapping air next to the body. This air acts to bar cold air trying to enter from the outside, and to prevent loss of heat generated by the bird. This northern ptarmigan tends to have longer feathers than its southern relative, the ruffed grouse; longer feathers have more overlap, increasing their insulative properties. In the winter, the Ptarmigan now before us would look quite different. Not only would her plumage be a pristine white, it would also be thicker. Birds of cold areas, like mammals, grow a thicker layer to cut the bite of winter.

Ptarmigan have also evolved a behavioural adaptation to reduce their winter stress. In places where snowdrifts accumulate, like the cranny where we saw the family of birds, ptarmigan take advantage of a natural coincidence. Snow provides the same trapped-air insulation as their feathers, and the ambient temperature in the centre of the drift is warmer than at the surface. By plunging into a snowdrift, the camouflaged ptarmigan escape some of the cold, and many of their predators.

"On our way I found a nest with four eggs. I heard ptarmigan flying by every half hour, which reminds me of back home."

Betty-Ann Betsedea - Canada
July 5

Another of the heat-wise adaptations the ptarmigan has developed is size. Larger, plumper birds are better able to retain their heat than are small birds because the core of their bodies is further away from the surface and the cold air.(4) No hummingbird would manage near the Kazan. Approximately sixty percent of arctic birds are pigeon-sized or larger, whereas in southern Ontario, the figure is only about thirty percent. Southern birds can afford to be less efficient because they have ready access to food. As food is the fuel burned to produce heat, a heat efficient bird will require less food to survive. This can be critical during periods when food is scarce; the capricious weather of the barrenlands is capable of locking away the high-energy insects and seeds under a sheet of ice.

The northern latitude of the Kazan places another major constraint on its nesting birds: time. For many species, the time period between arrival and departure is six to eight weeks. During our seven-week sojourn, we witnessed courtship and territorial behaviour and later found fully grown young preparing for migration. Early in the season, as we paddled past island colonies of Arctic terns, the birds would wheel and scream overhead, short-tempered with any creature that came too near. Some of the birds carried fish in their bills, courtship gifts to strengthen the bond with their mate. If we endured the warning screams and swoops to observe the activity at the colony, we could sometimes see the males placing the fish in the bills of the females, and displaying their black cap to intimidate any other male that approached too closely.

Within a few weeks, young terns, their foreheads showing white because of their incompletely developed caps, were flying expertly, ready to leave the Kazan. This restrictive time frame provides little opportunity to recover from a nesting failure. Most species simply have to wait for the next year, whereas a second set of eggs could be laid in the more southerly boreal forested region of Canada. There are many years when unseasonable storms destroy all nests in a region. The spring before our visit to the Kazan valley, an ice storm devastated numbers of the earliest returning small birds.

The effect of a lost breeding season can be a major population decline, from which it may take a species many years to recover. Some species, such as waterfowl, accommodate this contracted nesting season by courting and forming pairs while on migration, or by returning to the same site yearly. By conducting these preliminaries en route, valuable time is saved and eggs can be laid soon after reaching the nesting area. Except for the possibility of death from late spring storms, birds that hatch earlier have an advantage in that they are larger and stronger for fall migration.

"We went for a hike to the inuksuk across the ridge from our camp. Along the way we found many tree sparrow nests, some with hatchlings and some with eggs. This was exciting for me because it made me realize how much life really exists out on the barrens."

Ashley Wooten - United States

As we observed birds in the Kazan valley, we were often struck by the fact that nearly everything about them was designed to cope with their special tundra environment. During a walk, a muffled explosion of wings would precede something small and brown flying out from underfoot. If you bent down to examine the ground, the search often revealed a tiny cup of tightly woven grass, half-buried in the springy turf. Tucked inside might be four pale green eggs speckled with rust, no bigger than the end of a little finger - a tree sparrow nest. Such a find filled me with the same thrill I had experienced upon finding the empty shells of a robin's egg as a child. Little birds come out of those things! As a child, though, I had never seen eggs in the nest; the nests were always located in a tree, out of my reach. On the tundra, where there are few trees to provide homes, most birds have had to adapt to ground nesting. These adaptations usually involve some form of concealment. Arctic birds are not the most colourful in the avian world; most come in combinations of black, white and brown, and are nearly invisible against the backdrop of tundra. Camouflage is probably the most effective means of foiling predators. Even if the presence of a nest is revealed by the rapid departure of the incubating adult, finding the nest can be very difficult. Often they are hidden under tufts of grass or in the base of shrubs, or set right into the mosses of the tundra. For every nest found nearly immediately, there were half a dozen that were never found.

Although most sparrows and finches simply flee when the nest site is approached too closely, many shorebirds do elaborate distraction displays to lure intruders away. These displays include flopping and fluttering on the ground quite pathetically, or feigning a broken wing to entice a predator into pursuing the adult and leaving the nest alone. One semipalmated plover gave an award-winning performance near our camp one day. We had placed the tents in an area where our movements would not disturb her, but wanted to find the nest and mark it to ensure it would not be tread on accidentally. With three of us scouring the area, we nearly gave up to allow the poor bird some peace, when the four eggs were found, laid directly on the sand, speckled exactly like the nearby pebbles. We placed one of our orange sticks, usually used to mark archaeological features, in the sand near the nest, and studiously avoided the area until we left the next day, taking the stick with us. The eggs could so easily have been crushed, wasting all the efforts of the parents.

Birds are highly mobile creatures and therefore bird distribution is not constant. Distributional changes can occur for a variety of reasons. Climatic improvement over time may allow invasion of areas which were previously unsuitable for a species; migrational errors may introduce a species to a new area; loss of habitat through human alteration of the land may force a species to invade other areas. Species are sometimes introduced by man, both intentionally and accidentally, and rapidly adapt to and expand into the new location. The ability to fly provides birds with the ability to quickly relocate, for whatever reason. As the human population grows, human land requirements also increase. Every year, wildlife populations are placed under more pressure as their available habitat shrinks.

Arctic birds are not excepted from the influence of human development. The Arctic is rich in mineral resources, and is an important area for military defence. These activities, as well as pollution and the changes which may occur due to the Greenhouse Effect, all affect bird populations. Although arctic birds are superbly adapted to life in this area, many are functioning at their maximum stress level. Any additional stresses may be insupportable. For example, seabirds can keep themselves warm when feeding and resting on the Arctic Ocean, even in winter. If even a small patch of their feathers becomes matted with fuel oil, however, their heat retention capacity can be affected enough to result in their death. Birds have fixed upper limits of temperature regulation, and have adapted to work within the boundaries of the natural world, not a man-altered one.

If we accept that human land requirements preclude the preservation of all existing wilderness areas, it follows that we should attempt to save the areas of most significance to wildlife. The logical first step toward this goal is to identify important wildlife areas. This is difficult to do in southern Canada, where humans have already usurped much of the prime wildlife habitat, but it is still possible in the north. Although the Arctic is a vast and largely unoccupied area, not all of it is rich in wildlife. Large tracts of land are unused by wildlife, so that the areas like the barrens which are used, are essential. Many of the areas important to caribou have been identified but, except for a few high profile species (usually game) such as snow geese, the breeding distributions of most arctic-nesting birds are based on extrapolations from a few known sites. The Arctic, and particularly the barrenlands, has been little surveyed for its avifauna.

The design of breeding bird atlases, whereby actual breeding locations are mapped within a grid, defines a species range as units of area known to be used. Not only does this type of survey identify those areas which are important to each particular species, it also defines those areas which are used by many different species. Protection of these areas could provide the most conservation value per unit of land. As breeding bird atlas data are collected within a specific time framework, they lend themselves to comparison with data collected within a similar time period in the future. Such a comparison facilitates monitoring of a species, or an area, over time to assess any changes which may occur. Monitoring of protected populations is also necessary to determine whether the reserved area is large and diverse enough to effectively provide for the species, and to observe the effects of pollution and other influences which do not respect park boundaries.

One of the tasks facing the expedition was to survey the Kazan valley for its breeding birds. Similar surveys, using an amateur labour force, have been conducted around the world with excellent results.[5] The books produced provide current information on the status of the birds of a geopolitical region.

Birds lend themselves to these projects for a variety of reasons. Unlike the thousands of plant species which may be found in an area, the number of birds is usually a manageable 100 to 200 species. Because they are marked distinctively, make their presence known through song, and are largely active during the day, birds are relatively easy to observe and identify.[6] As a result, there is a large pool of capable birdwatchers who already record information on the birds they see. It is merely a matter of having birdwatchers incorporate atlas methodology into their birding activities.

> *"You would have thought that because the barrenlands have no trees, just little bushes, you'd be able to find birds really easily because they couldn't hide anywhere. But they could hide. It was hard to pick them out and keep an eye on them - to write or sketch anything for identification. They would just disappear into thin air, it seemed. They were everywhere, you just had to learn to look for them."*
>
> *Annie Roberts - Bahamas*

The difference between our project and its predecessors was that many of the expedition members had no previous knowledge of birds at all. To remedy this, we developed a crash course in bird identification and breeding bird atlas methodology, and slotted it in among the various other training sessions scheduled before our departure.

The classroom was filled with bewildered faces and the sound of rustling pages. Hands flipped back and forth through the pages of field guides as eyes scanned for some bird that resembled the image we had just projected on the screen. We were attempting to simulate field conditions by showing a bird for a few seconds only. Birds have a most annoying habit of moving as soon as you get your binoculars on them, so it helps if you can reduce the number of potential candidates quickly. Fortunately, in the Kazan valley, we expected only 66 species. Outside our classroom in Peterborough, Ontario, you can see 100 species on a good day. Our idea was to teach recognition of a bird as a member of one of the 15 families represented in the arctic, and then use the field guide to identify the bird to species. This is actually quite easy to do.

Loons (Family Gaviidae) are nothing like hawks (Accipitridae) which are nothing like sandpipers (Scolopacidae). Once you've slotted a bird into a family the number of species to choose from is narrowed down to as few as two or perhaps up to ten. The next thing to do is look for something distinctive about the bird: a band across its chest, a bright patch of coloured feathers, certain coloured legs, white on either side of its tail. With practice, you learn which features to look for to separate a particular bird from its closest look-alike. You also learn where to expect to find certain birds and from where to mentally exclude others. We always found water pipits on large rock outcrops, which seems to belie their name. Gray-cheeked thrushes were always nestled in the heart of an inland willow stand. After a while, you stop expecting to find them elsewhere, which helps rule out species.

Our objective was to record every bird for which we found breeding evidence [*see Appendix 1*] in a series of designated areas. In this case, the designated areas were every 10 km by 10 km grid square we passed through on our 500 kilometre route. This system enables the production of a map of the breeding range for each species. Each square in which a species was recorded as nesting is represented by a dot on the map. All the dots together give a real indication of the breeding range. This is an improvement on the "best-guess" technique used for the range maps in most field guides, in which the maps are based on a few real records with the rest of the range filled in hypothetically. Often these maps are quite accurate, but for some species, particularly those in the north, there is simply not enough information to provide reasonable guesses. The data collected during our Kazan River expedition will help rectify this situation.

One of the most difficult aspects of the bird survey was keeping track of which 10 km by 10 km square we were in. The squares are designated by a grid which is marked on all Canadian topographic maps. Each square has a unique code identifying it, and we used these to keep our records organized. Rivers, however, don't pay much attention to artificial grid systems, and we often found our course nicking the corner of a square for a hundred metres, or bending into a new square and then doubling back into one previously traversed. It was important to keep close watch on the maps to ensure that all records were assigned to the correct square.

A critical part of the breeding bird atlas system is to map only what is actually found, resisting the temptation to interpolate. It is very easy to start assuming, "Since we found willow ptarmigans over there, they must be nesting here as well". This sort of logic can quickly lead to a distortion of a breeding range. Similarly, it is vitally important to be absolutely certain that species identification is positive. Our policy was very much one of cautious recording, and in case of any doubt as to identification, a sighting was omitted. In this way, only definite records were included in our data.

The plaintive cry overhead seemed the loneliest sound ever uttered. Looking up, we saw a large bird circling. Even before registering the white tail with its ink-dipped tip, we knew it was a rough-legged hawk. Nothing else made that haunting sound. We had no time to search for a nest - the others were waiting for us at the canoes. I pulled out the data card we had started upon entering this square, and quickly scanned down the long columns of bird names printed on it until I came to the hawks. Next to rough-legged hawk I pencilled an "O" in the first of the four columns on the card. This was the only one of the 23 possible breeding codes that described what we had seen for this bird.

This lowest category, Observed, contains only the code "O" and is used when a species is seen but gives no indication that it is breeding nearby. Hawks, and other far-ranging species, often fall into this category because they travel long distances from their nest-sites to hunt. There was no telling where this Rough-leg had made its home. Even though Observed is a low category, it still provides useful information on which species are using an area. The three other categories and their hierarchically ranked codes reflect breeding evidence of increasing levels of certainty. When we transcribe our data onto maps, the three categories are distinguished by different symbols to aid the interpretation of the findings. As we left our solitary hawk, I wondered if, before we entered a new square, we would be able to upgrade the mark on the card to the second category, Possible, containing two codes. These codes are the lowest indication of breeding, and describe species that are seen in their typical nesting habitat, or a male in habitat singing to attract a mate. These two codes are ranked low because they define more an intention to nest in the area than an actual attempt.

Lunchtime. This cool grey day had gnawed away at our breakfast, and we pulled into shore earlier than usual for a bite to eat. Everyone was lounging, relaxed. There were no fish, or tent rings, or caribou, or birds to make any of us leap up, scrambling for cameras, binoculars, fishing rods or notebooks. We were content to merely let the time pass until the decision came to make way once more. Someone disappeared into the bushes, and immediately a wild and prehistoric trumpeting startled us into attentiveness. To me, this was a sound of remote and very wet marshes; I had forgotten to expect it here on the dry tundra. The others were captivated by the sound, having no idea what shape its maker would take. We waited patiently for the creature to emerge, but it stayed hidden until the shrubs were again disturbed by the return of the missing group member. Then two large, brown forms lifted into the air only ten metres away, and flew, long necks extended and legs dangling, a few metres to a ridge. There they stopped calling and settled down to graze. Birds in suitable habitat; "H" in the Possible category. I looked at my companions to gauge their reaction and to see if any of them knew what the birds were. Some were stunned, all were impressed by the sudden appearance of the sandhill cranes. There is something hauntingly primitive about these birds. Reminiscent of pterydactyls, they transport us back to a period when man was a less dominant player than he is today. For many on the expedition, the dramatic entrance of the cranes formed their first link with the bird world. They were strange, wonderful, but perhaps more importantly for the new birdwatchers, large, easily identified and slow-moving. The cranes were one of the first species to cooperate with novice attempts to focus binoculars on them, unlike the little brown sparrows which disappeared in a blur of wings and seemed impossible to identify. From this moment forward, sandhill cranes were our daily companions during the expedition.

> *" We have been making many sightings of breeding birds so far, but the one we saw today must be one of the rarest here: a pair of sandhill cranes with a nestling. Somehow, it sure doesn't look as pretty and graceful as the cranes of Japan. It also makes a terrible 'garooo-a-a-a-a' call."*
>
> *July 10*
>
> *"It seems that sandhill cranes are not so rare here as I had first thought. We have seen a lot since our first sighting - probably because they are so big and their calls so loud. And to think we were so excited when we saw our first pair."*
>
> *Boy Mow Chau - Singapore*
> *August 1*

Even with the cranes, however, we had no evidence that pairing had occurred - the type of evidence we needed to use the third category, Probable. It contains seven codes to cover those behaviours that signify the breeding process has begun, including defence of territory, courtship or copulation, and agitated behaviour. This in-between category is often useful when an observer has only a limited time to spend in an area. Birds often act suspiciously when disturbed, but don't resume normal behaviour until after the intruder leaves. If there is no time to lurk in hopes of catching a clearer indication of nesting, the Probable category must suffice.

The eight of us were making our way carefully toward the small inlet. The ground under foot was muddy and slick, tilled by the hooves of the caribou and irrigated by the water that streamed off their coats after crossing the inlet. We were heading for an archaeological site we had to map. It was an old Inuit camp with a few tent rings and several large meat caches. The shore of the other side of the inlet was thickly covered with willow, many of the branches draped with tufts of caribou hair from their moulting coats. As we pushed through the willows, several birds flitted into the central branches of the dense shrubs. They called "pink pink", perhaps to frighten us off, perhaps to reassure each other. We stood still, some of us crouching low in the willow to be less obtrusive. I "pished" a few times to try to draw the birds out. Pishing is an assortment of noises ranging from "shpishpishpish", to kissing the back of your hand, all of which often prove irresistible to nesting birds and mortally embarrassing to the non-birders in attendance. Instantly, three birds popped up to the top of the closest shrubs, all with the unmistakable head pattern of the white-crowned sparrow. Something about the way the bold black-and-white head stripes converge at the eyes gives them the look of punk rockers. A quick look through the willows and the ground in the area did not uncover the nest, so amid their incessant "pinking" we resolved to leave them alone. Their agitated calls allowed us to place a mark in the Probable category next to their name on the data card.

The final category defines those activities which indicate nesting has occurred. Because the thirteen codes within this category denote certainty, it is called Confirmed. Types of evidence include a distraction display to lure an intruder away from a nest, carrying nesting material or food, and finding a nest with eggs or young. It is desirable to use confirmed codes as often as possible to provide the most definitive breeding evidence. We always kept eyes and ears open for information which would allow us to upgrade a code to a higher level within a grid square.

The hill we were approaching was marked by a number circled with green on our topographic maps. A note on the back of the map informed us that this was an historically known peregrine falcon nest-site, perhaps noted by previous canoeists, or by a passing helicopter. Our task was to see if it was active. Peregrines are faithful to their nest-sites, often for years, so there was a good chance we would find birds. The adults are highly territorial, and defend their nest vociferously. Before visiting the area, we had developed a strategy to minimize disturbance of the birds should they be present. We would stay together as a knot of people to limit any disturbance. Upon reaching the hill we would search the cliff face, the most likely location of a nest, as quickly as possible, make notes and leave. If there were adult birds defending the area, we would restrict ourselves to five minutes on the hill and then leave, regardless of whether we had found the nest. At least we would know it was still an active site. So, with our game plan in hand, we began our approach.

Within five hundred metres of the hill, a pair of peregrines began crying and circling overhead. It was exhilarating to see them, their face-patterns clearly visible, their wildness permeating the air we breathed. Our careful plan was already obsolete - it would take us nearly five minutes to reach the hill over the spongy turf. We made the decision to continue, the choler of the adult birds increasing only when we began ascending the hill. Our sense of urgency mounted. At the crest, the search was rapidly made and rewarded nearly immediately. About one metre down from the cliff top, one of the three still-downy young falcons hissed at us. We circled to the front of the cliff face, quickly photographed the nest, made notes on its location, and hastily departed, the adult birds making warning swoops around us. Their cries dying away behind us, we were thrilled to have succeeded in our mission, and were satisfied that the adults had chastened us for only nine minutes in total; they didn't bother with us much as we headed away from the cliff. This confirmed peregrine falcon for this square. We marked "NY" (nest with young) on the card in the appropriate spot.

> *"After breakfast we biked up to see a couple of peregrine falcons and their nest. When we got close to the nest, the two adults started dive-bombing us. We took a closer look at the nest. The parents were going crazy, squawking and buzzing us. It sounded like sirens going off all around us. We managed to get within six metres of the nest and saw three chicks."*
>
> *Paul Clements - Jersey, Channel Islands*
> *August 9*

This particular sighting also meant completing two special raptor forms. As part of our bird survey work, we had agreed to provide detailed information on our observations of nesting raptors, or birds of prey. The Government of the Northwest Territories Department of Renewable Resources has been monitoring raptors since 1982; our data will contribute to their continuing program. Many raptor species suffered drastic population declines in the '60s and '70s when the pesticide DDT affected the development of egg shells and reduced reproductive success. Once DDT was banned in North America, raptor populations slowly began to recover, and are now monitored as an indication of environmental health. We recorded information on the exact location of the nest, the direction it faced, whether it was on a cliff or elsewhere, if it was a stick nest or just bare rock, how high it was, whether it was an active nest, and how many adults and young were observed. All these parameters help determine the health of the population and provide historical information for comparison with future studies. The procedure was very straightforward - any time we encountered raptors, or whenever we approached one of the previously documented sites marked on our maps, we attempted to obtain the information requested on the forms. This work emphasizes how effectively non-scientists can provide data of real value to the scientific community. Only basic observational skills, the ability to follow some simple recording methods and some sensitivity to the study's subject are required.

"A peregrine falcon can fly at incredible speeds. It lifts high, flapping quickly, folds its wings behind, and drops in a vertical dive at up to 360 km/hr. At this speed it hits a flying bird with its talons, stunning the victim and killing it easily by breaking its neck. It then rips the bird apart unless the victim is small enough prey to swallow whole.

"As we hiked closer, we spotted two birds flying above the ridge. Bruno took off quickly. I panted behind until I realized they were the peregrine falcons. The falcons dive at anything that threatens them. How they screeched and swooped over us. I didn't know if they'd attack or not, but a stony silence in their screeching meant they were diving. At each silent moment, I noticed myself cringe, my arm instinctively shielding my head. We knew we were close to a nest and went searching, conscious of finding it quickly, disturbing the pair as little as possible. Eventually, amid much noise and fury, I suddenly saw it, at eye level, a nest of dry twigs with three white fluffy balls, nonplussed at my presence. Their little black eyes stared back at me and their small dark beaks, hooked downwards, did not seem amused. I couldn't believe my discovery, shot a quick photo, counted the chicks carefully and left immediately."

Kassie Heath - Australia
August 8

I was always very grateful whenever we found a nest, or clearly saw a bird's markings, or witnessed a display of flight prowess. Personally, I was having a marvelous time with the birds of the Kazan valley - learning the different calls and songs, noting the plumage differences among individuals, guessing what we would find when we pulled the canoes up on a new area.

The frustrations of being a new birdwatcher, however, are still fresh in my mind. All birds sound the same, in fact you don't even hear most of them until you reach a certain interest level and learn to keep you ears tuned in. Birds move around a lot, so when you do actually find one, they usually fly off before you get your binoculars on them. It can all be very discouraging. We had spread the people with previous birdwatching experience among the four groups, and I was confident that our data would be complete and accurate.

I had set a secondary, personal goal for the expedition: I wanted one person, who had never thought about birds before, to develop enough interest to begin noticing the birds around their home area. I didn't expect to create a die-hard birder, I just wanted the Kazan experience to move someone to recognize birds as part of this world, and maybe have enough curiosity to identify the house sparrows and starlings which had previously gone unnoticed at home.

At the beginning of the expedition, most group members had sensory overload when suddenly set down in the middle of the tundra. Many of them were inexperienced campers and canoeists. Furthermore, the archaeological and treeline studies had much more concrete work schedules; there were specific survey areas with definite start and finish points marked on the maps. The bird surveys were on all the time, and were easily obscured by the immediacy of the other projects. For the first few days we saw little other than the "little brown jobs", usually flying away in an unidentifiable blur.

I tried to inspire interest by showing pictures of these elusive birds in the field guide, and passing along tidbits about behaviour or biology. The sandhill cranes helped. They were the first species to come out and say "TaDaaah!" and grab the attention of everyone in the group. Then, the next day, we flushed a short-eared owl, another commanding species, as we swept the ground for tent rings. As it flew off, everyone was able to see the black wrist marks and large neckless head which identified it. They would recognize it another time. Our daily encounters with Lapland longspurs made us all familiar with their black faces and chestnut napes. After a few weeks, people stopped saying "I saw a little bird that was brown and had streaks - what could that have been?", and started reporting "I saw a Savannah sparrow carrying food - do you have that yet?".

Although it was small perching birds like sparrows and longspurs that we encountered most frequently, it was our discovery of the more flamboyant species that sparked questions. Sleek and lovely parasitic jaegers sitting nonchalantly on the water as we paddled past five metres away. What do they parasitize? Tundra swans glinting white in the sunlight as they herded their babies between them. How old are swans when they learn to swim? The plaintive cry of the rough-legged hawk echoing over the remote rocky hills. What do they eat up here? These questions, and discussion of their answers, fostered an appreciation of the birds themselves, and also how the birds fit into the barrenlands ecosystem of the Kazan valley.

> *"I love doing the bird inventory. I wish I knew the birds a little better, especially their calls - it would make identification a whole lot easier and a lot more fun. But the birdwatching fascinates me - it is a newly discovered skill, and I am taking to it with relish."*
>
> *Sonia Mellor - Australia*

Parasitic jaegers are hunters, catching small mammals and birds for consumption. They also raid gull, tern and duck nests for eggs and young. Their name, however, derives from their habit of harassing terns until they drop or disgorge the catch they are carrying back to their nests. In fact, jaegers often accompany terns on migration, letting the terns do the fishing and then bullying them out of their catch. This is one particular strategy for survival.

Every species has evolved a breeding strategy as well. Birds which belong to the Order Passeriformes (including larks, ravens, thrushes, pipits, warblers and all the finches) are highly evolved and hatch what are known as altricial young. These babies are blind and naked at hatching and are completely dependent on their parents for protection and sustenance. Other birds, including ducks, geese and swans and the shorebirds, are precocial at hatching - that is, down-covered, sighted and mobile. Most of them can fend for themselves as soon as the down is dry, and can take to the water often on the day of hatching. Parents do attend precocial young, but usually provide food only to supplement what the young have procured for themselves. Adult tundra swans immerse their heads and necks under water, cropping aquatic plants and digging for the tender shoots and tubers. Young cygnets may be given some of this out-of-reach food by their parents, whose heads and necks are frequently stained with rust and brown from foraging for their vegetarian diet. This digging and pruning sometimes improves growing conditions for the plants, providing fertile grazing for swans returning the next season.

Species such as the rough-legged hawk, however, are unable to cultivate their food supply despite their dependence upon it. The major source of nourishment for these magnificent birds comes in a small brown package - lemmings. Lemming populations undergo cyclic fluctuations, reaching peak numbers and then crashing. This cycle seems to be linked to the supply of forage: when the food supply is healthy, the lemmings increase in numbers, until there are more lemmings than the dwindling forage can support. In this situation, lemmings migrate in search of new food supplies. Similarly, when lemmings are in good supply, rough-legged hawks enjoy favourable breeding conditions and their numbers increase. But, when lemming populations crash, the hawks must range farther afield to search for their scarce prey. Lack of food results in a decreased reproductive rate, and the hawk population generally follows that of the lemmings into decline. These long term cycles of high and low populations are also an integral part of the Kazan's arctic ecosystem.

We were crowded into a big kitchen tent, legs and elbows seemingly everywhere. It was a blustery, cold day, and the spray kicked up by Kazan Falls made the air damp. With less than a week to go before our arrival at Baker Lake, the chief birders from each group had gathered to go over our collective results, check for errors and develop a system to get the data into a final format before we left the north.

It was a tedious business. Every single record had to be perused, and the breeding code assessed for its appropriateness. "You put 'NB' for nest-building for willow ptarmigan. What did you see?" A pause while a notebook was consulted and the correct entry was located. "The bird was walking with some grass in its bill and put it under the edge of a dwarf birch." Good enough; it's not often seen but is the correct use of that code. "This greater white-fronted goose is marked 'FL' for fledged young. Shouldn't we use 'PY' as geese have precocial young?" Again the sighting was verified by checking notes, and the change of code agreed to.

It took hours to go over all the records, but there were moments of excitement, and envy, as new birds were revealed. "You saw a red-throated loon? Wow!" "Snow buntings and snowy owls in one day. That's not fair - we didn't see any at all . . ." I was conscious that our species list was quietly growing. Most of the sixty-two 10 km by 10 km squares we had traversed had been home to the same few species, the habitat being quite uniform on the barrenlands. But once in a while a few less common species turned up. As we looked at the cumulative results, it was exciting to realize that our group of amateurs had noted breeding evidence for most of the 66 species that had been predicted from the outset. Someone unzipped the door of the tent to let out some of the heat our bodies had generated. People shifted positions as they waited for the final figures tallied from the master card we had just compiled.

Our six weeks of effort had yielded 52 species, 31 of which were confirmed as breeding in the Kazan valley [*Appendix 2*]. Everyone was satisfied, spurred on to try to add a few more species in the next week. All the hours spent organizing notes and completing data cards were now transformed into something tangible and meaningful. Our meeting ended, and people went off to attend to various other tasks. There were three other scientific projects to address, suppers to be made and eaten, a rough-legged hawk nest to be viewed through the spotting scope from across the falls . .

One of the difficulties in visiting an area for the first time is that one has no impression of normality. It is only possible to see how things are during that time; there is no sense of context, no previous experience against which to compare current findings. Even the few papers that have been published on the birds of the Kazan area are scant help. They are also based upon short term observations, in an arctic ecosystem subject to significant fluctuations over time. To understand the status of any given species in one summer, one needs to know its status in the past. Thus, it is difficult at this point for us to make any extrapolations, other than in a general manner, on what our data may mean about the birdlife of the area.

According to current range maps, all of the expected species that we did not locate exist south of the area in which we travelled. They were included as expected species to remind us to keep watch for them should their range have expanded. It would appear that most of these species stay south of the treeline to have access to nesting sites in trees. As the tundra is interspersed with a myriad of lakes and ponds, we had expected to find large numbers of waterfowl taking advantage of this prime real estate. What we found instead was largely unoccupied waterfront property. We did record ducks and geese but, except for the large flocks of moulting Canada geese, we encountered this family only occasionally. We began proposing theories to explain their absence, and wondered if the year's severe drought in the prairies and southern Canada had affected waterfowl in some way.

It seems, however, that the Kazan and its vicinity is not a major breeding ground for waterfowl, largely because the ponds lack the weeds at their margins which provide cover for the nesting birds. Even loons, which nest on more exposed sites than other water birds, were in short supply. Although we recorded all four loon species, we saw very few individuals. The explanation for this remains elusive. Shorebird numbers also fell short of our expectations. Many plovers and sandpipers nest in the high arctic islands to the north of the barrens, but we were within range of semipalmated plovers. As I had seen 10,000 of them congregating on a beach during a fall migration stop in New Brunswick, I had expected to bump into them regularly on their breeding grounds. We did see them on many occasions, but well dispersed and never in the quantities I had envisioned. My fears of being unable to separate the very similar shorebird species proved largely unnecessary. I had plenty of time to study them between sightings.

Where numbers of some species seemed low, the exceptionally warm and dry weather of 1988 helped produce a bumper crop of other species. American tree, Savannah, white-crowned and Harris' sparrows were abundant, no doubt feasting on the vast quantity of mosquitoes and black flies available. Although these pests were an aggravation to us at times, they provide the principal source of protein for young birds. By the end of the summer, juvenile Lapland longspurs were so numerous on the dry heath that we were able to identify them intuitively, without looking at them closely.

Similarly, raptors in the Kazan valley appeared to be enjoying good health and nesting success. In areas where cliffs and rock outcrops provided suitable nesting habitat, the active nests had healthy looking nestlings. Although about half the previously used peregrine falcon nests were inactive in 1988, we often observed adult birds in the area. This indicates that breeding was likely occurring, despite the fact that we were unable to locate new nests. Peregrines tend to return to the same area annually, but varying food supplies or changes in the previous nest site can force them to relocate. We found one nest precariously perched on the edge of an eroding, sandy spire. Its one chick looked forlorn as if knowing that the nest would not survive long enough to support another generation.

There were a few surprising discoveries as well. One person found a few feathers from a northern flicker, a woodpecker well north of its usual range. It appeared that the bird had been killed by a snowy owl, but whether the kill was local or had been transported some distance remains a mystery. We found no other indication of flickers and, indeed, did not expect to. Although they feed extensively on the ground, they nest in holes excavated in trees, a scarce commodity on the barrens. Another species, although not unexpected, was a surprise in a contextual sense. When travelling in the Arctic, removed from civilization and with a very real sense of wilderness and isolation permeating the area, one expects the birds to be appropriately new and wild. It came as somewhat of a shock when a pair of suburbia's best-known avian residents burst out of a willow grove and flew away in terror. Obviously these were wild American robins, unlike the cheery fellows who probe our city lawns. It was very strange to think of them faring quite happily on the tundra where, to my mind, they didn't belong. This was another reminder of the adaptability of these feathered beings.

The results that are undocumented and, at the end of the expedition, still unknown, are the effects of the bird surveys on the expedition members. Before we parted company, there were many polite comments on how interesting the birds had been, and a few voiced intentions to buy field guides upon returning home. Certainly the barrenlands are a special place for all of us now, and if we were to hear of an ecological threat to the Kazan area, I'm sure we would be moved. But more than memories of a summer spent in an out-of-the-ordinary pursuit, I hope our expedition will represent the beginning of a greater awareness of the complexity of life on earth.

More than learning to identify a few species, I hope we learned to appreciate the intricacies of their existence, and the multitude of ways in which human life affects the balance of nature. Every country represented by one of our crew has a host of environmental problems - loss of species, dwindling natural areas, depletion of resources. Perhaps when they arrived home they were struck by the contrast to the quiet, harmonious system we lived in for seven weeks along the Kazan. Countries do not operate in isolation. The people of the world share one environment, and even the Kazan in its remoteness is affected by the changes occurring around the homes of man.

The migratory bird species of the Kazan are faced with alterations in their wintering grounds as well. The tropical rainforest that supports many arctic

species during the winter, and furnishes them with the energy to return to the Kazan valley, is disappearing at a rate of twenty hectares per minute.[7] If over-wintering success is decreased through lack of habitat, secure breeding grounds become even more vital for maintaining a healthy population. It is only by comparing what we have now to what we have at some point in the future that we can assess the effect of large-scale changes such as deforestation. It is becoming increasingly important to conduct projects such as bird atlases, to take stock of our natural wealth, and to accept the responsibility for their stewardship.

Twilight had deepened to dusk, the failing light reminding me that it was getting late in the season. In the last few days, sandpipers from the far north had been appearing along the Kazan, already influenced by the compulsion to move south. How could it be time so soon? I zipped my jacket up against the chill, tucking my binoculars inside; they were of no use in this light. My route back to camp was along the spine of an esker, a scar reminiscent of the great transfiguration wrought by the ice that created the Kazan valley. These natural roads pulled at my feet, beckoning me to follow into the heart of this emerging land. I started at the suddenness of the noise, and strained my eyes toward its source. The trumpetings rolled into each other in a practiced duet, a gradual crescendo as it approached. Finally, the sound of wind in feathers drew my eyes to the right spot. Against the backdrop of pewter water, the long-necked silhouettes of the two birds slipped by, seeming ghosts of cranes. Moments later, the camp silently materialized from the gloom where the Kazan had provided a place to stop. But the river continued, as we rested. It would guide us through this land for a few more days. Then we, like the birds, would return to winter homes.

Judith Kennedy - Canada
Expedition Biologist

Further Reading

Ehrlich, Paul R. and David S. Dobkin. 1988. *The Birder's Handbook: A Field Guide to the Natural History of North American Birds*. Simon and Schuster/Fireside Books, New York.

Godfrey, W. E. 1986 (revised edition) *The Birds of Canada*. National Museum of Natural Sciences, National Museums of Canada, Ottawa.

Hinterland Who's Who. Canadian Wildlife Service, Environment Canada. Information pamphlets on various species of wildlife.

Scott, Shirley L. (ed.). 1983 (2nd ed.). *Field Guide to the Birds of North America*. National Geographic Society, Washington, DC.

"I went up over the ridge behind camp, hoping to see the 'caribou' . . . there they were, thousands and thousands and thousands of them! Stretching as far as my eye could see. When I got close enough, about 20 m away, without scaring them off, I just had to sit down. The sight of so many caribou, freely ambling along over wide open spaces, limited only by sky and river, overwhelmed me. They looked so proud and strong, the bulls with their handsome antlers and the calves trotting beside their mothers, struggling to keep up. To see them scratch their ears, snort and grunt as they munched on the hummocks, so alive and so strong and so free, made me want to jump and scream in a frenzied passion! I closed my eyes and realized how beautiful and amazing and strange it was, all at the same time."

Eddy Chong - Singapore
July 27

ENCOUNTERS WITH WILDLIFE

The river is flat here, allowing us time to look around and to absorb whatever lies beyond our immediate watery surroundings. Its swift, directed momentum continues as it has done ever since it was released from the icy restraint of the glacier. Each year as the autumn days shrink against the approach of winter, the Kazan is once more halted in its progression, and its power arrested by the ice, a reminder that ultimately it is the ice which has granted it life. The ice shaped the land through which it flows, and each year it is the ice which gives over a part of its energy to the river, and nurtures it, for a few irrepressible weeks. During one of these short leases of life the river carries along with it our small canoes.

For several kilometres, the water moved over the land apparently without any guidance from the shore. The banks were indiscernible, any slope up from the river imperceptible. Yet such a slope must have existed, for the river is not capricious, but traditional; its route was designed by the glacier. We looked out over a table-flat land, on an ever-shifting, though dependable, scene that by degrees instilled a sense of acquaintance, recognition and comfort. We hoped for more: for insight and understanding, but these were yet early days for us.

> *"On the first day at Angikuni Lake, I thought the caribou tracks were car tracks, as my civilized mind could not think in terms of the barrens' wildlife."*
>
> *Angela Haas - United States*

The canoes were secured on the shore and we walked inland. In the distance, about two kilometres away, there was an elongated, low ridge running parallel to the course of the river. In a less flat land the ridge would not have been remarkable, but here its low hulk was the only relief for many miles. Moving slowly along the ridge was a small band of caribou, a mixture of cows, their new antlers of the season hardly visible, their new calves and around the periphery, a few smaller yearlings and an occasional huge bull with antlers already almost completely grown back.

If the Kazan valley belongs to one animal it is to the caribou, which travel back and forth across the land throughout their entire life. Theirs is a life of continual movement and change, of compliance with the moods and dictates of the seasons. The brief summer, lasting a maximum of two and a half months, is not an easy time for the caribou, for although food is plentiful, so are the insects. They appear in waves.

First the mosquitoes hatch from the myriad shallow lakes overlying the permafrost, lakes which dot the map to either side of the wide blue slash that is the Kazan. These are perfect hatcheries, still and shallow, and therefore soon warmed by the sun. The mosquitoes emerge for their brief life during which they plague the caribou in hordes, concentrating on their exposed soft muzzle, eyelids and lips.

A short time behind the mosquitoes come the blackflies, emerging from under stones in moving water, ready to feed on blood. The onslaught represents a significant drain on the reserves of the caribou who have a limited time in which to build up their body weight and health to see them through the seven months of winter. The caribou seek out high ground and snow patches whenever they can, where cool breezes keep the swarms at bay.

There are other, hidden pests. Warble flies have been incubating under the animal's skin to emerge in June through a burrowed hole. Nose-bot occupies and reproduces in the nasal passages and migrates deeper into the respiratory tract, causing intense irritation and often respiratory interference.

Through our bug-repellant headgear we watched the twitching flanks and the distracted head-shaking. Caribou are not equipped with a fly-whisk tail as are many of the ungulates of the African plains, and since the ridge offered scant food they had soon to return to the damper tundra to feed. There they must endure the incessant piercing, biting and scraping attacks of the insect swarms, since any prolonged respite would not come until autumn.

We took out our notebooks and recorded what we saw. One of our tasks was to record all the mammals we encountered on our journey. Mammal populations - like birds - are not static, in number or distribution, and the information we collected contributes to a larger data bank which will allow many questions to be answered relating to the status and possible vulnerability of the animals in this area. By simultaneously monitoring the distribution and abundance of mammals, and information about the individuals, their habitat and behaviour, we can gain an insight into their present ecology, and begin to learn about the colonization process which took place in the Kazan valley after the glaciers retreated.

There are many questions to which such data might be applied. What mammals live in the Kazan valley, and how many of them? Which part of the tundra do they utilize, and with which other animals do they share this space? Is there any unique aspect of the tundra which is used by an animal and thereby constitutes an essential part of its life history and survival potential? In what respects do the animals compete, and to what extent? Is one species out-competing another? Are the animals currently living in the Kazan valley fully adapted to their environment, and if not, then are they continuing to adapt? What adaptations do they show? Do some species adapt more quickly than others? What sort of human activities are likely to interfere with the lives of the animals? How would such interference be manifested? Assuming that any significant interference constitutes a disturbance of the internal balances of the system, how could we help to restore the overall balance? Is it necessary that we should, or will a new natural balance be achieved? Where will this be, and how long will it take to develop and stabilize? How will those (human) communities which still hunt on the land be affected in the interim, particularly if animal numbers and traditional movements change and become less predictable?

There are a number of parochial, specific questions to which the Kazan data will contribute. The Kaminuriak herd, to which the caribou we encountered in the Kazan region belong, has been monitored for many years. Any additional information about their whereabouts or movements, the size and the composition of the groupings is valuable. Wolves are the main predators of the caribou, so information on their distribution, numbers and activities contributes to the overall ecological picture and may help to explain observed changes in caribou numbers or movements. Only a few years ago, muskoxen were considered to be under threat of disappearing from this area, and in 1927 a reserve was set up for their protection along part of the Thelon River to the west of the Kazan. Observations of muskoxen in the expedition area are therefore useful in assessing the current population.

There may be other uses to which such baseline data may be put in the future. There are many examples of data being applied to problems different from those for which they were originally collected. It is important to use every opportunity to gather information about the nature and the mechanisms of changes in natural systems. Our questions cannot be answered on the strength of one season's data from the Kazan, but neither can they ever be answered without amassing such basic data. The starting point of our research must be the basic numbers and distribution details on which we can begin our descriptions and quantification of populations, and then we can go on to quantify any changes and ultimately to build predictive mathematical models of the process.

We recorded all mammal sightings, including wolves, arctic hares, arctic foxes, and the smaller mammals such as voles and lemmings. Some of them are herbivores and some carnivores, some predators and some prey. Recording everything we saw provided valuable background information on many mammals. Experienced researchers use such information in assessing the state of health of the tundra and its value: ecological, aesthetic and economic.

Although geographically remote, the barrenlands are under ever-increasing pressure of development from outside as the search for resources, especially minerals, expands to feed demand. It is important that we know exactly the relative ecological and development potential of different areas, so that we can base future decisions on sound information. There is already competition for land use - between the native hunters and developers - and increasing tourism will add a further dimension to the problem. It is essential, therefore, that we gather the necessary information on which to be able to formulate land-use policies, which will minimize the extent and duration of any damage to the ecosystem on which so much wildlife and so many people depend.

Within the limitations of assessing a caribou herd at eye level, we recorded in our notebooks the details of the caribou on the ridge. How many? Males? Females? Calves? Other young? The colour and condition of their coats. The size of their antlers. Which animals had antlers? Were they in velvet? Where were the animals and what were they doing? Were they utilizing a particular feature of their habitat? What time of day was it? In which direction were they moving and where was the wind coming from? It didn't take long, though the animals were quite a way off. It was not always possible to answer all the questions, but binoculars and a telescope helped. Above all, we had to be sure of the accuracy of our notes, for much depends on them. It would have been interesting to linger a while, and to watch the behaviour of the animals, to acquaint ourselves with them more closely, but we had several kilometres yet to travel before we set up camp. So we returned to our canoes.

Before the end of the day we located one of the spruce groves we were to study, and made our camp close by. The grove was small, but dense, and the trees provided a perfect haven for the mosquitoes which had begun to emerge in the warmth of the previous few days. We began to understand the miserable plight of the caribou we had seen the day before. It was very difficult to concentrate there. The work was slow because of the tangle of branches, and frequent interruptions in measuring and note-taking to swat the mosquitoes and clear the air immediately around you for a time. It never worked, of course. There were plenty more where they came from.

During a coughing fit, brought about by the ingestion of one of the swarm, a member of the group bent over to rid herself of the intruder and found a small ball of grass stems just under a willow bush. It was the winter nest of a lemming, about twelve centimetres across, and very solidly made from entwined stems of grasses which had long since died and turned yellow. There was a small hole built into one side of the nest, about three centimetres across. This led into the fur-lined interior. The nest was robust enough to withstand handling without being deformed, but it was not possible to pick it up since the lemming had secured it to the surrounding vegetation. In several places six to ten centimetres away from the nest were small, distinct, tidy piles of droppings. In common with nearly all animals, lemmings do not soil their nests, but create convenient latrines. When the area around the nest has several such piles, the lemmings move out and build another nest.

During the summer lemmings excavate systems of tunnels in the moss, and these include fur-lined chambers which serve as nurseries. The lemming is aided in its digging by a long, flat nail on each of its fore-paw thumbs, and broad, strong claws on the lateral three digits of its front paws. The animals may be seen above the surface, especially in peak years of their population cycles. On one occasion as we packed up camp, a lemming shot out from between two stones and ran across the area where the cooks had been standing to prepare breakfast. It did not appear alarmed, rather it seemed purposeful and busy. Even in the summer the animals are active throughout the 24-hour day and much of the time is spent in feeding. Such a small animal has very few reserves and cannot afford to be caught out by a storm which may confine it underground for several days at a time. There were many such glimpses during the summer.

Mammal sightings became more numerous downstream of Yathkyed Lake. The increase began with an explosion of caribou which we observed one day while camped on the east shore of the river where it leaves Yathkyed. Our tents were erected close to the shore and behind them a ridge rose gradually up to its heathery top. Behind this lay a higher ridge which formed the horizon and, when viewed from the lake, had appeared to be covered with spruce trees. Now, seeing it at closer quarters from the ridge in front, it became clear that the "forest" was one of antlers.

Caribou in their thousands progressed southwards along the ridge and overflowed down its flanks and across the valley floor between that and the ridge on which we stood. The animals were not milling about, but were clearly driven by some common instinctive urge. The herd moved smoothly, without apparent haste, but without any doubt as to their intention. Individuals were moving quite rapidly, judged from concentrating on a single animal to see how quickly it covered the ground. Caribou have long legs, an adaptation for a more efficient running gait common in the deer family, and their easy, apparently slow, pace is deceptive.

At intervals, animals stopped to graze the nutritious young tissue of the herbs growing under the tougher heaths. Others did not stop but seemingly poured around those who were grazing and continued on their way. The feeding animals were not left behind because hundreds more caribou came up onto the ridge from behind, or filed along its sides and around the shores of the three elongated ponds situated in the valley bottom. Wherever we looked the tundra was speckled with their relatively pale bodies. So this was one of the huge summer herds, we thought. In fact it was only a fraction of one part of the entire Kaminuriak herd (estimated at 300,000). Cows drop their calves in the area around Kaminuriak Lake to the east of the Kazan and the majority winter several hundred miles south of where we now stood.

Individuals looked shabby with their dishevelled coats. None looked sleek yet, as they would before the winter storms arrived; the lines of the ribs were still clearly visible on a few of the animals. They sported a range of colours, which is normal at that time of year as the light winter coats are shed, revealing new dark brown coats underneath. The new calves were the palest, although the yearlings were not very dark. The main body of the herd was composed of cows with their new young of the year, but in this group there was a higher proportion of bulls.

Bulls occurred in groups within the herd, though the sexes were not thoroughly intermixed, and it was the antlers of one of these groups which had caught our attention as we canoed across the last kilometre of Yathkyed. The cows were as yet without obvious antlers, but those of the bulls were held high. Although they were all well-grown, a few stood out from the rest by virtue of the enormous thickness of their tines and the well-developed palms between. They were still growing, too, apparently, for they were still in velvet, though not for very much longer. These were senior bulls indeed.

"We saw 20,000 caribou! These last few days have been magical, a dream, but just now I have returned from seeing maybe 20,000 caribou and I am breathless. We hiked up a ridge at the northeast end of Yathkyed Lake, about 700 metres from our camp, where we'd spotted maybe 10,000 caribou the day before. In the distance, white specks were sunlit on the hillside - many caribou. They were walking towards us, following their age-old paths, caribou highways of the tundra. We walked to the next ridge and by the time we arrived, the first of the herd were there on the windward side. We stole silently along the ground, crouching low, towards a rock pile, where we could sit semi-hidden. The caribou passed within 10 to 15 metres, observing us a little warily but content enough to graze and scratch around us. The first herd was quite small and after they'd passed we spread ourselves out 20 to 50 metres apart, lying prone on the grass, the damp and cold seeping up

through my thin trousers. We were intensely excited but silent, heartbeats loud in our ears. We needed to wait only half an hour before the next, much larger group moved around us. Their hooves clicked as they walked past, like a thousand knitting needles. This time we were amongst them - caribou passing two to five metres away, lying down just metres in front to rest or nibble at low shrubs and grassy tussocks, their lips moving quickly as they foraged. They scratched themselves: hind legs with antlers, foreheads with hind legs, rubbing necks and jaws with each other. Occasionally they'd glance our way with a snort, a huff or a grunt. They trotted along, stopped to graze, trotted again - scatty behaviour. We were the creatures that stared in awe. They seemed very nonchalant. Eventually, the cold and the damp were too much. As the herd moved to the north of us, swinging around our little ridge, we walked back to camp knowing they would be here for at least another day. Thousands and thousands of white specks peppered the distant hills. There were caribou as far as we could see."

Kassie Heath - Australia
July 27

Watching so many animals move across the tundra in front of us made us understand how the caribou trails, which we had seen since we began our journey, had come to exist. On many occasions we had looked across the vast tundra from a slightly elevated vantage point and had seen, picked out by a low sun in pink or gold, or shaded in charcoal as a child uses a soft pencil to shade the relief of tree bark, hundreds of closely-packed, anastomosing trails. They were aligned like lanes on an infinitely wide highway, and this is exactly what they were: the expressways of the tundra. Most of the deeply rutted trails were little wider than the hooves which made them, so that it was impossible for a human to walk normally along them.

We saw such trails in many places on the tundra. In some cases they ran parallel to ridge systems, either skirting the base of the ridge or climbing it and then descending the far end, while in other places they ran at right angles to a ridge, cutting across its end. In these cases the trails often led directly to the water's edge, at a place where the caribou habitually cross the Kazan.

During the spring migration of the caribou, north from their winter quarters within the forest margins to their calving grounds, and their return journey in the autumn, they must make many river crossings. Early in the season the river is still frozen. But such an advantage is not always available, and the caribou then seek crossing places where the river is calm, shallow or narrow.

These relatively safe places have become traditional crossing points, and they testify to the huge numbers of animals using them each year. We came across a number of such places as we carried out our surveys along the river, but one was particularly memorable. Here, the wet peaty trails made their way across a low, wide mossy shore and intercepted the river at right angles. They were spaced a metre or two apart where they entered the river, though they were denser a few metres inland. In total these tracks stretched for about two kilometres along the shoreline.

As we walked over the area it was not difficult to envisage the herds which had crossed there, for as we glanced across at the opposite shore we could see, about a hundred metres or so downstream, that the vegetation was a less vibrant green. It looked tired from having so much life trampled out of it each year. With such a short growing season the repeated battering had taken its toll and the caribou's point of exit from the river was permanently marked.

This spot also brought home to us what strong swimmers the caribou must be. The river here was calm but swift, and about three hundred metres wide. It was no easy swim to reach the far shore while losing only a hundred metres ground. Their swimming power was later demonstrated to us directly as we sat quietly on a shallow, pebbly shore awaiting the return of an archaeological survey party.

A group of cows and two calves appeared on top of the opposite bank almost directly across from where we sat. We made no noise or movement and they continued down the steeply sloping, heathery bank to the water's edge. They barely hesitated before wading in and finally launching themselves into the cold, swift current. Two cows went first, followed by a calf and then by another cow and finally a calf entered a few metres in the rear.

They made amazing speed, and barely seemed to lose any ground at all. The river was about a hundred metres wide at this point and as they swam the cows made soft, encouraging, grunting noises to their calves, and occasionally turned their heads to check on them. The first calf kept his position in the line of cows, but the trailing youngster fell behind further and further.

Its mother reached our shore only about seventy metres downstream from where we sat, while her calf was still in midstream. As other animals finished their swim and shook themselves, they started to feed. The mother whose calf was still in the water shook herself but remained on the shore facing the river, watching her calf and occasionally calling to it. After a minute or so she turned her back on the river and resumed grazing.

The calf was carried several metres downstream, but continued swimming valiantly. It appeared to us he was not going to reach the shore, but a few minutes later he rejoined the group and began to feed as well. By this time the others were making their way slowly up the slope behind us, grazing as they went. The mother did not look up. It seemed she had recognized that her calf was safely across even before it emerged from the water.

Before we had seen the apparently boundless herd of animals on the northeast shore of Yathkyed Lake, it had been hard to imagine how the trails could be formed. Now it was clear. Caribou instinctively follow the animal in front, to the extent of using its hoof print as a guide to where it should place its own foot. Such behaviour conserves energy through the harsh winter. By using the trail which has already been cut through the snow, all but the leading animal save significant amounts of energy. Anyone who has walked through even knee-deep snow will readily appreciate the advantage. During their spring and late autumn migrations, and their meanderings in winter to find food, caribou seek routes which take them along exposed ridges or across windswept frozen lakes. Such places offer the easiest walking conditions, and thus conserve their energy.

Caribou feet are a superb, composite adaptation for both soft snow and hard rock or icy surfaces, in addition to being effective paddles which propel them surely through swift currents. Each foot is formed of two half-hooves which run protectively around the outside and across the top of the soft foot. As the animal transfers its weight onto the foot, the two halves of the hoof splay and the soft foot spreads, offering an increased surface area of contact, reducing the pressure and the amount the foot will sink in. At the same time, the soft underside of the foot affords a degree of traction, but most of the grip is provided by the sharp lower edges of the hooves which act like skate blades against the ice.

We remained at our camp on the north shore of Yathkyed Lake for over two days, waiting for our re-supply of food. Caribou continued to pour through the valley and across the ridges, rambling and eating the fresh summer growth as they went. As the sun dropped low in the sky and gilded the tundra, the caribou began, a few at a time, to lie down for the brief "night" and to chew their cud. Other animals filed past them but eventually the entire herd was motionless for a spell, except for the movement of jaws.

The outlines of the bulls' antlers were illuminated as the sun passed through their velvet. The scene was restful, and a stark contrast to the incessant movement of the day, but the glow of the sun belied the keen wind tossing the platinum heads of the bog cotton. This was no time for us to linger. We emerged, stiff and cold, from the small depression behind the tussock which had concealed us for hours, and returned to our tents, leaving the caribou to the cold.

The temperature was probably around freezing when we left. As the caribou lay on the cold ground to rest for a few hours on their southerly journey, they were insulated by a dense coat which is composed of two types of hair. Next to the skin is the truly insulating layer of fine, dense hair or, strictly, underfur, which traps warm air and reduces the heat lost from the warm body of the animal. The tough outer, shiny, guard hairs grow through this and lay flat along its surface to create a windproof and waterproof outer shell, thus retaining the warmth of the trapped air beneath. The guard hairs are hollow, trapping more air.

The efficiency of the guard hairs at enclosing the insulating fur is clearly demonstrated when caribou swim. The air in their coats provides buoyancy and they swim high in the water. The guard hairs keep the underfur dry as they swim, and a quick shake when they emerge sheds the water clinging to them. It is no accident that caribou skins were highly valued for clothing by the people who once inhabited the Kazan valley. Although there may be a substitute for caribou meat as food, there is nothing to replace the natural properties of caribou skin and fur.

We saw many more bands of caribou roaming across the tundra as we proceeded down the river over the next few days. None of the groups was as big as the Yathkyed herd, but we saw hundreds of animals at a time. The composition of the groups appeared to change as we moved north and east from Yathkyed Lake, with an apparent increase in the proportion of bulls. This was possibly so, for the sexes do not always move together. They spend the winter in their own groups of females with calves and a few yearlings, and other groups consisting of immature animals and bulls. There is no rivalry between adult males except during the rut which occurs on the autumn journey south to their winter quarters. After the rut is complete, the bulls are tolerant again of one another's company and remain associated throughout the winter and spring.

When the days lengthen the pregnant cows move off first and make their northerly journey to their calving grounds on the tundra. They are accompanied at first by their calves of the previous year, but before the cow drops her new calf she will reject her yearling and the young animals then form their own groups which remain on the periphery of the calving cows.

The bulls begin their northward migration commonly a week or two, but sometimes up to a month, behind that of the cows, and do not rejoin them until after the cows have given birth, when they move off to feed. It was at this point that we met the animals. They were roaming the Kazan valley to feed on the nutritious young growth of the tundra sedges and herbs, no doubt a welcome change from the tough, scant diet of lichens on which the animals subsist through the winter. We were witnessing the groups rejoining for the summer.

"Our canoes carried us down swift currents, over great expanses of lake, through wind-tossed waves, down the Kazan River between banks sometimes high and rocky, other times low and green. But canoeing was a means, not an end. We were there to explore the river valley. The canoe was our vehicle for this journey through an emerging land."

"I jotted down the data as someone called out the figures - spruce tree species, height, diameter, presence of male and female cones, mature or immature tree, and the shape that the tree had grown. As the figures accumulated they created a picture of a typical spruce tree - 150 cm tall, 10 cm diameter, sprouted as a juvenile layer from the roots of a mature tree. Only in the most northern grove had we discovered seedlings: sprouted from cones that almost all mature trees produce but by which means they seldom reproduce.

Kassie Heath - Australia

"Inuksuit were cairns of stone, sometimes placed strategically in lines near water-crossings to allow families to drive caribou into the water where hunters waited in their kayaks to spear them.

"Inuksuit were often a metre or more high, decorated with chunks of vegetation or bird's wings to make them appear more life-like. Caribou spooked by a line of such apparitions would avoid crossing it, hurrying on to meet their fate."

"We had a close encounter with a family of three muskox - a bull, a cow and calf. They came right down to the river to drink about eight metres away from us. It was really exciting. Then they ran away. All three muskoxen ran straight past within about a metre. It was a great feeling of fright and excitement as the large beasts ran past. The bull then stopped and raised his head to make a grunt-like sound, his black coat blew in the wind as he stood poised on the top of a small hill. The sight was truly magnificent."

Paul Clements - Jersey, Channel Islands
July 23

"We walked out across the tundra and lay down in a line to watch the herd. They seemed completely unaware of us, and walked up to just below the knoll that we were lying on. I was actually there with caribou two metres away from me, lying down, chewing, and grunting. I could smell them, a funny musty but sweet smell. This is not the sort of thing I ever expected to see in real life."

Simon Cremer - England
July 27

"There were two major finds: Unguluk's camp as reported by Tyrrell in his geographical account of the Kazan, where we could discern three tent rings, two parallel lines of stones (maybe to support kayaks) and hearths and maybe a meat cache.

"Our second great find was at the end of the transect where a hunting region became evident - two tent rings very separated and one marked with inuksuk, two hunting blinds separated and facing in different directions, many meat caches with built up walls and a bone crushing region where marrow was extracted.

"This was someone's home. It makes the whole place seem so much more alive."

Kassie Heath - Australia
July 8

"Today was tremendously interesting. We actually got to see more than just the surface of a tent ring. By unearthing it - it almost felt like we were a part of it, living and using it."

Leslie Mack-Mumford - Toronto, Ontario
July 23

"We dug one square metre at a time with trowels. It took me five hours to complete three squares digging to a depth of 10 cm. As most of the artifacts are found in the thin layer of peat, we only dug until we hit gravel. This is usually not more than 20 cm at this particular site. So far the items we found were mostly caribou and muskox bones and teeth."

Boy Mow Chau - Singapore
July 20

"The river had cast its spell. Each one of us knew we would carry in our hearts a special feeling toward that river for the rest of our lives. Some spoke of challenges they had met, others of the wildlife they had encountered, others of the new friendships cemented on the journey, and others of the land through which they had travelled. For everyone it was a time of connection to the river, a river they all hoped would somehow be preserved for time immemorial."

Caribou are shadowed closely in their migrations by wolves, their main predator. Wolves are not strictly either an arctic or a migratory species, but they have a long association with caribou, in northern regions, which stretches back into the ice ages. The association has been refined to the point that, on the tundra of central and eastern Canada, the distribution of wolves extends to cover the area described by the annual movement of the caribou.

Wolves are adaptable and are equally at home in the forests, where the caribou spend the winter, and on the open tundra. It is not the cold which controls their distribution, but the availability of prey. They congregate around the caribou calving grounds and can often be seen trailing a wandering herd. There is no question that a caribou is much swifter than a wolf; its long, thin legs are perfectly designed for a quick, efficient pace. All else being equal it can easily outrun a wolf. The predator must therefore use other tactics to bring down its prey.

Wolves overcome their lack of speed by hunting in packs and organizing the group as humans might organize a military operation. Each member of the pack has its own job, and together they follow a plan of action. A common method is for a part of the pack to run the prey into an ambush of their hidden companions. Initially there is careful selection of the prey. Wolves will watch a herd and identify any weak, slow or injured animals, and these will then be hunted. A wolf is able to judge whether or not it can outpace an individual caribou. The ambush technique is effective against the swiftest animals, but wolves will scatter a herd to separate a weak animal before finally running it down.

"We were happily chatting away about nothing in particular in our little tent, all lying slightly down slope so that we faced the river, when I told everyone to be quiet. I had been listening to a clattering noise travelling near to the tent for a few seconds. I said, rather casually, 'there's caribou out there' - thinking I might have a look, or maybe not bother as there would always be more caribou to see - when a mother and calf came bolting around the left side of our tent and started clattering up over the huge boulders. Rapidly, in a crouched hunting position aimed at the caribou, came a fluffy, sleek, black-as-night arctic wolf. I was the first to see the beautiful beast, as he fled across the front of our tent - not two metres from our heads. After I rather rapidly and excitedly informed the others, the three of us made enough noise by sitting upright in our sleeping bags to get a better view, that the wolf stopped dead in its tracks, looked at us, looked at the caribou, decided we were too great an "unknown" element and hastily retreated back to where he had come from. But the wolf wasn't going to

give up his well-earned meal of caribou that easily. He moved about six metres and then turned back to us and dropped to the ground - basically out of our sight! We were all thoroughly excited. Inside the tent, all was a flurry of exclamations, of grabbing for cameras and of slight fear of a rather hungry, wild animal lurking outside with rather sharp teeth and us in our flimsy little tent - between that beast and its dinner. We then all stuck our heads out of the window which was too much for the wolf, who reluctantly started to move away, never going more than ten paces without stopping to look back. I can't describe nor contain the excitement I feel."

Sonia Mellor - Australia
July 30

"We were all in the tent, when suddenly we heard the sound of running feet outside, accompanied by a nervous snorting noise. We all looked out of the tent and to our amazement saw a caribou and her calf looking very anxiously towards us. They clambered and clattered over the rocks and onto the other side of the ridge - obviously greatly disturbed by something. We did not have to wait long for an answer, as all of a sudden we heard a "woof" - like a rush of wind blowing through a tunnel. An arctic wolf! No more than five metres away. It was jet black and as it panted I noticed immediately how red its big tongue was. It was salivating at the mouth, and as it puffed, it seemed to smile and I could see its white, white teeth glowing in the evening light. It crouched, bunching its shoulders, moving slowly and steadily towards the caribou. Its eyes so intense, the glint so steely, so cold."

Richard Wilson - Scotland
July 30

Although caribou have reasonably good sight and can see approaching objects from a considerable distance, they rely on their sense of smell for final confirmation of identity. Thus a wolf is able to stalk closely if he remains downwind. Even if the caribou catch his scent they will not necessarily alert the rest of the herd and run away. They sense from the posture and behaviour of the wolf whether or not he is hunting. We witnessed this remarkable fact on a calm, warm day as we walked inland to record the details of a small caribou herd which was grazing to the west of the river.

The caribou were walking south along the ridge top and along its flanks. Following the base of the ridge was a shallow depression about ten metres across in which shrubby willow was growing. In front of this lay the flat heath-covered tundra on which we stood, about 200 metres distant. Caribou were moving around, disappearing into the depression and reappearing to continue up onto the ridge. As we watched we picked out two wolves, trotting in a relaxed attitude alongside one another, away from us, and positioned well within the throng of caribou.

Both wolves were of considerable size, and their coats, one a pale fawn and the other a dark silvery grey, appeared thick and shaggy, exaggerating their true width as we looked at them from the rear. They were moving across the flat tundra about twenty metres from the depression and their progress was slightly faster than that of the caribou, who were stopping to feed, totally unperturbed by the archenemies in their midst. Caribou were all around the wolves, though they remained about five metres from them, and it was not possible that the animals downwind had not caught their scent. Clearly, the behaviour of the two wolves did not contain the tell-tale signs that they were on the alert and hunting for food.

It is true that wolves are to be found accompanying almost all reasonably sized caribou herds, and that large numbers associate closely with the cows and calves on the traditional calving grounds. However, not all wolves that trail the pregnant females as they leave their winter quarters will arrive at the calving grounds with them, for the female wolves must themselves give birth. This usually happens a week or two before the caribou drop their young. Furthermore, the wolves must prepare a den two or three weeks before their young are born. The open tundra around the calving grounds is not suitable for this purpose and so wolves will often drop behind the caribou and rejoin them later. This is obviously an advantage for the caribou whose most vulnerable time is around calving.

When choosing a place to locate their den, wolves seek dry areas close to a supply of drinking water, with light soil in which they can excavate the necessary tunnels and birth chamber. Shallow permafrost and wet substrate preclude the general tundra surface from use, and wolves denning in such areas will utilize the sandy banks of an esker. The entrances to such dens are relatively easy to find, but others are less so.

On one occasion after making camp we were made aware of an occupied den by the yelping of the pups. From the camp there was no outward sign of its existence. It was located in a rocky bank of the river where the boulders were piled up into a steep promontory, and the entrances were the natural holes between the boulders. Oblivious to this subterranean lair, one of our group walked over its roof. She was greatly surprised by a view of a wolf on the far side. She returned to the camp in great haste, and, with a mixture of fear, agitation, excitement and joy in her voice, delivered a deluge of words which we understood to mean that she thought she had just seen a dog, or a wolf, or a fox, or something. Indeed she had. Her description left us in no doubt and one of us followed her back up to the promontory to take down the animal's particulars for our notebooks. The animal was no longer to be seen, and it was not until later that evening that we heard the pups inside and realized that she had discovered an active wolf den.

"I saw a wolf on a boulder shore. This was my first experience with the wolf because my country is a modern commercial centre. I was excited and scared. The wolf stared at me with shining eyes from ten metres. I guessed its shock was more or less as mine. It went away after about 20 seconds but it was a very long time for the two of us. We found that it had a den underneath the boulders and heard the voices of the young. The wolf came in and out of its den to watch. I will never forget this contact with the wolf."

Choi Siu Ping - Hong Kong
July 12

From then until we left the camp, we cautiously observed a hundred metre wide no-man's land between our camp and the wolves' territory. At intervals, an adult animal would appear from behind the piled up boulders and stand on top observing us. We only ever saw adults: either the pups were too young to come out or the parents were enforcing their confinement because of the perceived potential threat to their safety. Despite their obvious watchfulness and presumed anxiety, they showed no signs of aggression over the time we were camped there. It was clear that so long as we maintained the empty border zone between our respective areas, we were not considered a serious threat.

When we finally left the camp and canoed past the boulder promontory, there were no obvious prey remains or any other littering of the bank. Had we not landed there, we would have overlooked the den.

The adult that arrived at the lookout on the roof of the den displayed a motley coat composed of a range of hair colour from silvery grey through pale fawn, light brown and dark grey. The outline of the body was smooth, if a little thin, and it was not clear from across our border whether the summer moult had been completed or not. However, wolf colours are extremely variable, and although the pelage of the adult was patchy, its smooth appearance suggested that this was the new summer coat. Its texture was in marked contrast to that of the broad, shaggy animal we had seen as one of two moving amongst the caribou herd a few days before, yet the coat of the companion was much finer, if not exactly sleek.

All mammals shed old, worn fur and replace it with new. The coat of the shaggy wolf we saw was probably its dense, winter fur not yet grown out. In mammals which inhabit areas where there are seasonal changes in their environment, the density, colour and texture of their seasonal coats may change accordingly. Thus, many arctic mammals grow a white, dense coat in readiness for the winter snow conditions, and regrow a darker, thinner summer coat when the time comes.

Arctic hares grow a white coat in winter which is much denser than their summer pelage. White hair lacks the usual pigment (melanin) which gives mammals their brownish colouration, and as a result the individual hairs are hollow. So animals such as the arctic hare gain additional insulative properties from their white fur over and above that of the increase in density.

Arctic hares were recorded on a number of occasions as we surveyed higher, exposed, rocky ground. There the animals gain a long view over the tundra and can get the earliest possible warning of approaching danger such as a wolf or some aerial predator. They rely on sound to a great extent and their long ears are always held erect, but in a land where the wind is often ferocious, having a clear view confers a significant advantage. Arctic hares are huge, weighing up to five or six kilograms and measuring between sixty and sixty-five centimetres.

Typically they are seen sitting up on their large hind feet watching. Apparently this pose is very comfortable since they hold it for several minutes if they do not feel directly threatened. As we walked over the high ground we caught glimpses of a hare bounding off fitfully, stopping every few metres to peer out from behind a perched boulder to check on our movements. When they arrived at what they considered a safe distance, they would halt, not necessarily out of view, and take up their alert pose, sitting upright on their haunches.

All the hares we saw were in their summer coats, which grow through in June. At this latitude this is a silvery grey colour over the back, the ears are black and the feet and tail a silky white. They are not well camouflaged; the silvery hue of the back contrasts greatly with the matte grey of the exposed rock on which they move about, and even when only their ears are visible, pointing up like two small triangles from behind a rock, they advertise the animal's presence very effectively.

The winter coat of the arctic fox allows the animal to continue its activities outside in the usual winter temperatures. The insulative properties of its fur are such that it can remain comfortable while at rest in temperatures down to minus forty. Of the arctic foxes we saw, most had their summer coat, usually complete by July. They look as though they have been hastily daubed with a nutty brown paint in an attempt to cover up their natural colour, which resembles that of old paper. The back, tail and outer-legs are a fresh brown colour and their bellies, lower flanks and inside-legs a yellowish white.

An interesting sighting of a fox occurred during the steel grey light of a late evening paddle. In the middle of the sea-like Yathkyed Lake, all was calm and quiet. The water took on the colour of mercury, and rippled lazily as the boats moved slowly through it. As the canoes passed beneath high, rocky ramparts which bounded a tiny island, the quiet was interrupted by the liquid calls of several snow buntings, which appeared like night watchmen, one on each of the highest boulders atop the ramparts, and proclaimed our presence to the other inhabitants of this small insular community.

An Arctic fox displayed typical curiosity in coming up onto the boulders and satisfying himself of the cause of the disturbance. He stood looking down on us, attentive, as our canoes passed below him. Reassured that we would not be landing, he turned and trotted back inland. He was in a tawny summer coat, with some darker brown coloration in patches. How had he got there? The island was several kilometres from the mainland, so presumably he must have walked across the lake when it was frozen during the winter. It was another reminder that we were just travelling through, catching glimpses and making short acquaintance with the land in one of its most tolerant moods. The fox on the island must know other moods.

We were fortunate to see ermine along the river; fortunate not because they are particularly scarce but because they are small and move extremely quickly, darting about under cover most of the time. We had camped on a wonderfully comfortable, aromatic bed of herbs, interspersed with just enough moss to provide a soft mattress under the tent. The bank along this stretch of the river consisted of a broad wall of huge rounded rocks each up to a metre or more across. Those towards the top of the bank were apparently stable, and had been for some considerable time, judging from the size of the variously coloured lichen colonies that encrusted them. The size and thickness of the colonies decreased towards the water, where the rocks were smaller and had been repeatedly pushed up by the winter ice.

As we left the camp one morning to do an archaeological survey, a long, slender, cinnamon-coloured animal shot out from a hole between two boulders, caught sight of us and froze. His tail was tipped with black. After a few seconds, during which both he and we remained totally motionless, he seemed to regain some confidence and took two or three more short steps before stopping and watching us some more. Our shapes were probably unfamiliar to him, for he then rose up on bent hind legs and sniffed the air to help him in his identification. In doing so he showed off his pale apricot belly. He was handsomely turned out in his summer livery, now some three months old, and he looked sleek and healthy. It is no doubt significant that this was one of the places where we saw lemmings on the tundra. The ermine had found a good hunting ground, and would probably remain there throughout the winter to hunt beneath the snow, by then wearing his pure white coat but still with the black-tipped tail.

Two of the most exotic inhabitants of the Kazan region are undoubtedly the grizzly bear and the muskox. Grizzlies occur eastward from Alaska, where they are most numerous, across the Yukon and into the Northwest Territories. The Kazan valley is approaching their eastern limit, though the animals are regularly sighted there. We saw no bears, but shortly after pitching our tents beside a sandy beach to the west of Yathkyed Lake we discovered a long trail of fresh prints in the wet sand less than a hundred metres away.

The prints were quite distinct, showing clearly the imprint from the ball of the foot and from the toe pads. One print, twenty-four centimetres long and eleven across, included five sharp incisions into the sand made by the long claws. The prints belonged to a female and a cub, which at that time of year, in July, would be six or seven months old. They were fresh, judging from the sharp detail we were able to record, which then faded over the next twelve hours.

The two sets of tracks led along the damp sand and through a high willow thicket growing at the outlet of a small stream. They emerged again on the far side of the willows and continued across another sandy area before disappearing . . . where? The imprints were quite even, that is to say their depth was more or less uniform and there was no sign of the heels being dug in, which would have suggested that the animals had hurried away, alarmed. Instead it seems that they may have moved off when disturbed by the distant sounds of our approach, if they had not already done so before we arrived in the area.

Grizzlies roam over the tundra during the summer, feeding on any available fresh, succulent plant growth, but they are omnivorous animals and will take carrion or kill their own prey. This may include ground squirrels, of which there is no shortage, hares, foxes, caribou and even muskoxen. In winter the bears make a den into which they retire until the lengthening days of spring arrive and the temperature increases. During this period they remain in the den and sleep, but this is not a true hibernation and they may wake at intervals and move a little. During their confinement they neither eat nor defecate.

We stayed at the "grizzly" camp for several days, busying ourselves with archaeological work, and finally left as the cold amber flood of another dawn poured onto the beach. We were slowly becoming attuned to the living rhythms of the land; some, like the dawn, were precise, others were more fickle, like the wind. We lived with them both and tried to work with them as a part of a finely tuned instrument in which each component carries out its task at the prescribed time, when its motion will have the desired effect and will not interfere with, but will complement, the other parts of the mechanism. This was the way to survive all the moods of the land, the way the animals lived, and how the Inuit had survived. For us, on occasion, it meant leaving at dawn in order to dovetail our canoeing with the wilder nature of the daytime winds. We were not used to working with natural cues and we had to learn what they might be and which were important. Through generations of necessity the Inuit who lived on this land became extremely finely adjusted to their natural environment; we went as far as we could along the same trail during our very brief stay.

The mammals of the Kazan region have adapted to their world over time, but none is more at home in this environment than the muskox. In fact, its total distribution is restricted to certain northern tundra and polar desert regions. Its heavy, lethargic appearance brings to mind the large lumbering animals of the mammoth and woolly rhinoceros world, and its ancestors once shared their wintry home with these now extinct animals. The muskox we now know has remained largely unchanged for about two million years. It survived in refuges during the last ice extension, moving back into northern central Canada from central and eastern parts of the United States.

After some time on the river, though not initially, when we searched in willow thickets for firewood, we began to find little straggly clumps of wool clinging to twigs. Sometimes the clumps were mere wisps of gossamer-fine, tawny wool, sometimes they were considerable bunches, several centimetres long, and some were fuscous, slightly more coarse, streamers. The wool was a sign that muskoxen had been through and were moulting their massively thick underfur. The prospect of seeing these primordial-looking creatures, representing two million years of arctic survival, was exciting, and not entirely expected.

Muskox numbers on the Canadian mainland, as elsewhere, had been seriously reduced by hunting in the 19th century. In 1927 the Thelon Game Sanctuary was set up along part of the Thelon River, lying to the west of the Kazan, in order to protect the animals and to allow their numbers to recover.

This they had done, but in 1982 when a small group of canoeists made their way down the Kazan there was no sign of any muskox along their route.

So the woolly evidence was exciting on two counts; as well as raising our hopes of seeing the animals, it pointed to an eastward extension of their range. Since muskoxen do not make very lengthy expeditions over the tundra in the course of their search for suitable feeding grounds, the suggestion was that numbers had increased.

For several days we continued to see the discarded wool draped in the lower reaches of willow bushes, and occasionally we came across sizeable willows which had been severely damaged. The thick stems had been twisted round and round so that the bark was split and shredded, to reveal the fresh yellow wood beneath. And the stems were doubled over so that the top of the bush came close to the ground. In some places we found dead, grey willow bushes deformed in exactly the same way, and obviously many years dead. Muskoxen were the most likely culprits here. They are the only animal strong enough to handle live wood of that girth that would have any apparent reason to do so, possibly to bring the young, tender leaves into easy reach for browsing.

"Every day since Padlerjuaq has been an excellent animal-watching day. Yesterday we had a muskox at the camp. Today, at the mouth of the Kunwak River - where I caught my first fish of the whole trip - we saw a wonderful looking muskox, which I believe had some intention of crossing the river but which decided against it when he saw us (the ugly, smelly old lot that we are) on the opposite bank. I went crawling through the tundra and had a wonderful view of him nibbling on birch shrubs - pulling the leaves off with his huge tongue, then, when he had a mouthful, rolling his head back to swallow the contents. He was a delight to watch. I could have stayed there all day. When we first spotted him, he stopped next to a rock and spread-eagled his legs to scratch his belly on a rock. He looked rather self-satisfied."

Sonia Mellor - Australia
August 8

Then one day our constant vigil of the land was rewarded. Away in the distance, among the endless hummocks of the tundra vegetation, was a group of rather larger dark brown hunches. Our curiosity aroused, we landed our canoes and slowly and carefully made our way towards them. They were a good kilometre or two away from us, at the base of a bare-topped ridge, and moving towards us at a slow ambling pace with their heads down, feeding. The ground was very uneven, with firm grassy tussocks on which dwarf birch was growing separated by small depressions or narrow channels just large enough to tread in.

Our progress was deliberately slow, though we could have done little else, for we kept one eye on the ground and one on the muskoxen the whole time. At first they seemed unaware of our approach; we walked hunched over to reduce our height and soften our shape. The muskoxen continued to graze on the lush, damp grasses and sedges which grew under foot in the depressions. Occasionally one or another animal would look up and we would crouch motionless for a while until it resumed feeding. Eventually one or two individuals noticed our approach, and they began to look up more frequently and for longer periods. However, their reaction was not one of alarm, but of uncertainty, in what we might be, not about what they should do, for the group continued to move gradually in our direction.

Then, when we were within about 300 metres of the animals, one of them suddenly appeared to decide that enough was enough, and that he should let the others know his feelings. The group then stopped for some time and watched us, still in the loosely arranged, sweeping, line they had maintained throughout. While their gaze was on us we did not move, but after some time they began to relax a little and one by one resumed grazing. The watchful bull continued to break from his feeding more frequently than the others, and we crept forward by small degrees whenever he put his head down. In this way we came to within 200 metres of them, but at this point they started, turned and retreated a few metres. We remained still, and shortly they described a semi-circle and began to graze again. It was clear that we could get no closer, so we quietly settled down to watch them about their daily routine.

They were not entirely at ease with our presence, as their grazing led them across in front of us at some distance, and then around in a large arc and away. Though they were not relaxed with us, they were clearly at home in the general surroundings. There were three males in the group, distinguished by the huge bony plates which cover the forehead and almost touch in the front middle line, four females, whose bony plates are less massive and separated by a tell-tale tuft of hair, and three calves. The calves moved among the adults, which always remained at the flanks of the group.

In mid-summer, the bulls, many of which have been solitary since the beginning of summer, rejoin the groups of females and immature animals in readiness for the sparring and competitive head clashes of the rut which takes place from mid-August to September. The horns are an important part of the ritual, with the bull having the largest and most darkly stained horns being most likely to win out against lesser bulls, and thus gain the right to breed the harem with which he has associated.

Calves are born in late April and early May when there is still plenty of snow about, and although they have a thick covering of underfur, the long silky outer coat that gives the muskox its characteristic shaggy, prehistoric appearance does not develop fully for another two years. The calf is well insulated from the cold, but must guard against wet, and the young animals will disappear under the skirts of their mothers during showers of rain or wet snow.

When the guard hairs do grow through they are impressively long, especially those on the chin. Together with the thick underfur which exaggerates the real body size, they form an efficient insulative layer which allows the muskoxen to stand inactive for hours, or longer, in the face of the worst winter blizzards. Any snow settling on them, as it often does because of their almost torpid behaviour, does not melt. This apparent lethargy is vital to their survival. By minimizing their energy requirements, muskoxen reduce the amount of food intake they require.

Studies show that muskoxen alternately eat and rest throughout the day, switching between the two at roughly ninety-minute intervals. They both graze and browse, taking roots rich in carbohydrates, fresh herbaceous forage which is rich in nutrients, and browsing willow. Muskox are able to utilize the upland areas, and this they tend to do in late autumn and winter, when the ridges are blown free of snow and the heath species such as bilberry and crowberry are exposed.

The Kazan valley is an extreme place in winter. The climate is very harsh, with temperatures plummeting far below freezing, often accompanied by winds of gale strength. Snow accumulates in hollows and sheltered places, but the ridges are blown clear. This is most important to the muskoxen, whose range is known to be controlled at least in part by the degree of snow accumulation. They clear light crusts of snow away with their feet - their horns are ill-placed for this use - and heavier crusts are cracked by a hefty blow delivered with the muzzle before being scraped away again with the feet. This imposes fairly strict limitations on the snow conditions which they can tackle, and added to this is the simple problem of mobility. In deep snow muskoxen cannot move easily because their long arctic evolution has equipped them with short legs to reduce heat loss. They are thus restricted in winter to the rather more windswept areas, particularly the uplands, where snow does not accumulate. The Kazan valley offers a significant range of windswept uplands rich in browse.

Their thick coat belies the bulk of the animal, which stands much less than two metres high at the shoulders; usually it is about the height of a man's chest. The bulky insulative layer serves to surround the small appendages such as the tail and ears. The disadvantage, however, is that their hearing is reduced, though still of some use. The muskox has been compensated for this in evolution by keen senses of smell and sight, despite the apparently tiny eyes, another illusion created by the mass of hair around them.

The bony eye sockets of the skull are set into the end of tube-like protrusions which position the eyes clear of the horns and fur, and so increase the animal's angle of vision.

Muskoxen can move at a considerable speed when pressed, and if possible will head uphill. Their wide, acute vision and sense of smell is vital for them in sensing their only serious predator, the wolf. Wolves can take a healthy muskox if they find it alone, but muskoxen have a very efficient method of protecting themselves in groups. At the approach of a predator, they form a semi-circle or a full circle, closely packed together with the younger animals behind them in the centre, and face outwards toward the predator. As long as all animals keep the ranks closed, this is an effective defence against even packs of wolves, who may circle for over a day before giving up. This is a very specific response, apparently requiring an advanced level of co-ordination between individuals, which presumably has evolved over the long period that wolves and muskoxen, and their respective ancestors, have lived alongside one another.

Eventually the muskoxen moved away and were joined by another male who appeared on top of the ridge and slowly made his way down to them. We returned to our canoes, our thoughts captivated by the venerable looking creatures and their apparent easy familiarity with the wilful, harsh environment of the Kazan.

"We were stalking a small herd of muskox, which was grazing on the opposite shore of the river. It was very exciting to see such wild, woolly animals - munching out on succulent willow branches. Every so often, they looked up quizzically, as if to say 'I'm sure I heard something unusual out there.' I would freeze, crouching down quickly behind a rock, and wait until the muskox had decided to eat some more grass. The muskox was accompanied by another fine looking bull, and these two gentlemen each appeared to be accompanied by four cows and two calves. Quite a clan gathering. I was so close I could hear them snorting, and smell them even across the river."

Richard Wilson - Scotland
July 27

It was a long way from Angikuni Lake down the Kazan River. We were carried through the unfamiliar scenes and spirit of the barrenlands. Through occasional glimpses of some of its inhabitants, and the quiet observation of others, a small proportion of their secrets were revealed to us. Through these insights we gained a degree of familiarity and understanding. We were brought more closely in tune with the barrens, and adapted to a degree. We have come to understand the qualities which are necessary for survival there.

The short time it took to familiarize ourselves with the mammals is itself instructive, for it is a reflection of the small number and limited range of species which have so far accomplished the necessary extent of adaptation to the environment. Not only is the number of species low, but the few which do have a key to survival are representative of a very small number of animal types. Of a total of ten mammalian orders and thirty-six families existing in Canada, only four orders and eight families are represented in the Kazan valley. Within this, there is variation in the number and diversity of species in each group. For example, the mouse family is represented by numerous species of voles and lemmings, and there are three species representing the dog family (red fox, arctic fox and wolf). There is, however, only one squirrel-type animal (the arctic ground squirrel) and one member of the deer family (the caribou). This raises all kinds of evolutionary questions and heightens our awareness of the active, selective (i.e. non-general) nature of adaptive responses.

It also brings home to us how relatively simple this emerging land is, when compared with those long-lived, diverse and complex ecosystems of the world, such as the coral reefs and the tropical forests, which have been evolving uninterrupted by major climatic change for millions of years. These ancient ecosystems exhibit a bewildering array of plant and animal types, and of interactions between the two, and such diversity confers stability.

This is not to say that they are immutable, for ecological stability is a dynamic state. The living elements that make up the rain forests continue to evolve, but it is something of a fine-tuning process. Any disturbance to a part of such a complex system will have the domino effect inevitable when elements of a whole are interconnected and perform according to the behaviour of their neighbours. But there is a very important difference between the effect of a disturbance in such a diverse, complex system and one in the Kazan River region where the ecosystem is less complex. In the first case, the effects of any perturbation may be dissipated in many different directions, through the many channels of the ecosystem, and hence their effect on any one element of the system is likely to be reduced and less severe. But in an emerging, developing and relatively simple system built on fewer working parts, there are fewer routes among which the effects may be shared. There, with only a few species making up the total system, the consequences of any disturbance (we might say imbalance) tend to be more concentrated, and therefore have a relatively larger effect, since there are fewer parts to take up some of the strain and to buffer the system against extremes.

The Kazan valley system is a unique heritage, rich in cultural and ecological lessons. Its inherent beauty and the sense of profound quietude it confers, despite the often wild nature of its weather, are things to be treasured. The area allows us the privilege of glimpsing a wonderful process of natural growth and development as it occurs. There are few places left of which the same is true. Left to itself, the barrenlands will continue to follow the natural course of increasing complexity and diversity of evolution. This major scheme of things is at present beyond our control, and its time scale of a different order of magnitude from our own. But even within the very small time scale in which we operate, we can have profound effects on the course of that succession, and its final destination.

Within our own time frame we must continue to monitor the relatively short-term, small-scale changes and attempt to understand further the interaction between the land and its inhabitants. In this way we can hope to avoid causing damage through unnatural disruptions to this fragile and vulnerable land as it emerges from the ice, moves towards maturity and finally comes of age.

Jane Claricoates - England
Expedition Ecologist

Further Reading

Calef, G. 1981. *Caribou and the Barrenlands*. Canadian Arctic Resources Committee and Firefly Books Ltd.

Chernov, Y. I. 1985. *The Living Tundra.* Studies in Polar Research, Cambridge University Press, Cambridge.

Lauritzen, P. 1983. *Oil and Amulets.* Breakwater Books.

Marchand, P. J. 1987. *Life in the Cold.* University Press of New England.

Pruitt, W.O. 1983. *Wild Harmony.* Western Producer Prairie Books, Saskatoon.

"Something struck me then (about the herd of caribou), and I couldn't quite place it till we were all sitting on the beach, huddled around a dwindling fire, later in the evening. How strange that I should be standing at the very same inuksuk that pointed the way to this rich hunting ground, where once an Inuk stood with spear or bow and arrow in hand, looking out from the very same place with his hand shading his eyes from the sun. He would have seen the very same river, shimmering blue in the warm summer light, and looked across the tundra to see the same masses of caribou that I now see. A smile must have etched his sun-tanned, lined skin with the knowledge that soon there would be food for him and his family. I stand on the same spot now, with a smile on my two-month summered face and a camera in hand, happy with the knowledge of what I have seen. I opened my eyes once more and there they were, not a dream, but a link to that long ago time that has brought me here. I stood there a while longer, happy in my thoughts of a time past."

Eddy Chong - Singapore
July 27

- HUNTERS - LIFE FROM THE LAND

On July 6 we lay exhausted on a grassy bank below the third cascade. The mile-long portage behind us was a memory of sounds - the rushing cascades, the alarmed cry of a tern, and the sound of our own footsteps squelching through wet tundra, and then crunching over a nine-month-old snowdrift on the final descent to the river bank. Bent double under our burdens the whole way, we had only a lemming's view of the land.

Tyrrell, the geologist, did not record his personal impressions of the same portage that he walked on August 26, 1894. His attention is diverted by the inhabitants of the area: forty-two Caribou Inuit were camped in seven tents below the portage. A hunter called Pasamut who had just returned from trading at Marble Island, some 400 kilometres east of here, drew a map to help guide Tyrrell towards a short cut to Hudson Bay. He also supplied the white man's expedition with 100 pounds of dried meat - the bounty of the land. Some people were cutting tent poles nearby from one of the northernmost spruce groves on the Kazan River. They also hunted caribou which, at the end of the summer, were now drifting south towards the forest. Such a large encampment could easily be supported by the caribou at this time of year, with plenty of dried and cached meat to spare for the winter.

The years around the turn of the century were among the best years for the Caribou Inuit. They were well equipped to live off the land with its plentiful supply of game in this remote interior wilderness. At the same time, they participated in extensive trading with the whaling ships, Hudson's Bay Company traders, and Inuit on the coasts to the north and east. Unfortunately, these years of plenty did not last. Today, Inuit elders in Baker Lake, Arviat, and Whale Cove remember the 1940s and 1950s as years of hardship and starvation. Thousands of years of human settlement on the river reflect much the same kind of ebb and flow of fortune.

The land has been reclaimed, on a somewhat precarious basis, for the living, since the final collapse of the continental ice sheet here 7000 years ago. Interdependent communities of plants, animals, and people have learned to live in this environment and cope with its uncertainties. The adaptation of people to the barrenlands was not a slow, steady, continuous process. Rather a series of inroads were made by small groups of unrelated people from both the boreal forest to the south and the arctic coast. Fragments of their past are still visible in the stone tools they left behind.

The Kazan River swept around us, carving out the sand and gravel we stood upon. It conveyed such a powerful feeling of motion that it was difficult to stare at the current and stand still at the same time. Rocks lay embedded in the sand at our feet, forming a circle. The land at our backs rose evenly to the horizon several hundred metres away, forming a ridge of gravel which loomed over this tiny site like a wave. Here, in this small area of calm, a family lived for a while, hunting and fishing nearby. Now marked only by a circle of rocks that once secured the caribou skin walls of a conical tent, there was no indication of when hunters had last been here. The site was modest, barely visible above the encroaching lichens.

Startled by a sudden shout, I looked up at the ridge where a solitary figure shouted again - "I think you should take a look up here". Expectantly, I left the ring and ran up the slope feeling as though I had just been informed that there were unopened Christmas presents waiting. The others from the ring accompanied me over the ridge.

This new landscape had a dry, ancient feel to it. Before us, a gravel surface stretched away like a roadbed prior to paving. Its sharply defined edges fell steeply on either side to the green muskeg below. This remarkable feature was unlike anything we had seen previously. It was obviously a natural formation - perhaps an underwater bar formed in the shallows of the frigid Tyrrell Sea at the end of the ice age. Even more remarkable in this strange context was a trail of smooth cobbles - a line about 30 metres long that began and ended abruptly, like the fossilized track of an imaginary bird which had skittered across the surface of the ridge. The fist-sized cobbles were just big enough to stand out from thousands of randomly occurring water-worn rocks in the gravel.

Around this line were hundreds of curved, thin flakes of white and pink translucent quartzite - the debris left behind by people making stone tools using a skill called knapping. The stone flakes may have been on the surface for hundreds or even thousands of years. Almost no vegetation grew on this stony, bone-dry slope to cover them. We followed the rocks to another line of cobbles set at an oblique angle to the first. These mysterious features were reminiscent of inuksuit formations.[8] Inuksuit were cairns of stone, sometimes placed strategically in lines near water-crossings to allow families to drive caribou into the water where hunters waited in their kayaks to spear them.

Inuksuit were often a metre or more high, decorated with chunks of vegetation or bird's wings to make them appear more life-like. Caribou spooked by a line of such apparitions would avoid crossing it, hurrying on to meet their fate. The small rock lines which we encountered seemed at first unlikely to achieve the same effect. On the other hand, they managed to hold our attention for several hours while we recorded them and their associated knapping stations. It was not inconceivable that these unnatural formations aroused mixed feelings of curiosity and apprehension in caribou on the ridge top looking for a place to cross the river. Inuit elders from the community of Baker Lake later assured us that the line of stones would easily deflect a herd of caribou towards the river, and waiting ambush.

The cobble formations appeared without warning and we did not find anything like them again. They seemed to be a fixture of the unusual landforms of the area. I was to learn during the course of the summer that there are few simple categories or typical forms in the archaeological record of the Kazan. Paddling down the river, we saw the country change, and with it, the character of the sites. People living in a rewarding but unpredictable land learned to be flexible. It is this record of life in the interior, of adaptation to an unforgiving environment, of resilience and sometimes failure, that gives the Kazan valley its distinctive heritage. The expedition was documenting fragments of this past - the camps, hunting grounds, and work places that people used seasonally, and sometimes from year to year, as experience and opportunity allowed.

The inland Inuit relied heavily on the caribou which migrated from the Manitoba forests northwards into their homeland each spring. The people were descendants of Inuit living along the Hudson Bay coast in the early nineteenth century. These coastal people were themselves part of the Thule cultural tradition - a sea-mammal hunting culture which spread from Alaska eastwards across the Canadian arctic as far as Greenland about a thousand years ago. The Inuit who came to live on the west coast of Hudson Bay hunted whale, seal and walrus on the ice and open water of this inland sea. In fall, however, they looked inland to the caribou that flowed southwards to the forest in vast numbers each year.

The first people to move to the interior for more than one or two seasons of hunting did so in the 1830s and 40s when declining numbers of caribou and an expanding human population prompted a search for game further afield. This search most likely brought them up the Maguse River to Yathkyed Lake. Some of the Caribou Inuit societies which developed and expanded after this interior movement, like the Paatlirmiut who continued to live along the Maguse

River and the east side of Yathkyed Lake, maintained ties with the coast where they hunted seal and walrus during the summer. Other societies, like the Ahiarmiut and the Harvaqtormiut who lived on different sections of the Kazan River, lost contact with the sea. They came to focus exclusively on terrestrial species, caribou and muskox in particular.

The bulk of caribou hunting was accomplished in the fall ". . . when the animals are returning to the forests, fat and newly moulted, with shiny, soft hair."(9) The carcasses were stockpiled for winter under mounds of rocks. Musk-ox were also a source of meat for the Caribou Inuit, particularly in winter when caribou were scarce. Gregarious and relatively sedentary compared to caribou, muskoxen were a seasonally important and reliable species in the diet of the Caribou Inuit during the 19th century when there were large numbers in the southern Keewatin. An increase in the market value of muskox robes and competition from American whaling vessels forced the Hudson's Bay Company to raise their prices in the 1870s, providing incentive for Inuit to bring hides to Churchill for trade in the winter when the hides were in prime condition.(10)

The inland dwelling Caribou Inuit flourished between 1880 and 1915 while game was plentiful. In 1915, and over the next ten years, the caribou were devastated by fluctuating winter temperatures which caused a layer of ice to encrust their lichen forage. The herds that people depended on each year did not return to the tundra, or were greatly reduced, with disastrous consequences for the Inuit. Muskox were no longer available as a back-up, having been decimated by hunting. It is estimated that two thirds of the Inuit died during this terrible period, reducing the population to about 500 from 1500.(11)

The legacy of this great famine was to make the Inuit more dependent on the outside world. Trapping became more important to the local economy. At the same time, the Hudson's Bay Company was looking north of the tree-line to the Arctic as a last bastion for bartering, its traditional means of business, which was being displaced by a preference for cash further south.(12) Arctic fox, useless as a food item, was the commodity on which this fur trade was built in the 20th century among the Caribou Inuit and which provided the means for their acquiring such staples as guns, ammunition, flour, sugar, tobacco, tea, and knives.

Although the Inuit economy was precarious, based as it was on the whims of European and American fashion, at least some Inuit viewed the situation as a good deal for themselves: "The kabluna [white men] are fools to give us so much merchandise for foxes which we cannot use, and which we can trap as many as we want in winter."(13) The winter fox-trapping season, however, was a lean and unforgiving one. Caribou meat, acquired and cached in the autumn, continued to be the critical source of sustenance during the winter. The Inuit were still largely dependent on the caribou, despite frequent periods of hardship and shortage, until their evacuation and resettlement in new communities like Baker Lake and Arviat (then, Eskimo Point), during the 1950s and 1960s.

"We usually spread out in a line and walked along the transect looking for anything of interest. Wood is very uncommon in the area. We often found little bits of rock, which at first we all walked over. Andrew kept pointing them out to us and we were slowly able to train our eyes to recognize the minute things as well as the more obvious, the tent rings and meat caches. At first, I didn't have any idea what I was looking for. I couldn't believe Andrew was picking up all these bits of rock and saying 'this is a flake'. We kept handing him all these little bits of rock and he kept saying 'no, this isn't anything of use'. Eventually, we trained ourselves to recognize flakes and cores and that's when it became interesting.

"Once, we were walking along and it was really boring because we couldn't find anything. We were just about to turn back when somebody found a copper pot. Then we found these long pieces of wood. Any wood was really exciting to find because we knew it had come from a distant place where there were trees. One piece of wood was a bit shorter than the rest and both ends were slightly wider than the middle part. Ian said 'wow, this is a kayak paddle.' Then, of course, everything made sense. The other, long thin pieces of wood were used for the hull of a kayak."

Leslie Mack-Mumford - Canada

Long before the Caribou Inuit adopted their year-round way of life on the tundra, before even the Thule Inuit began their eastward trek from Alaska, the Keewatin was home to thousands of caribou and muskox. The herds provided sustenance for hunters who visited the area seasonally, pursuing large game northwards into the void left by the retreating continental glacier. These earliest hunters are called Palaeoindians. They occupied sites on the Dubawnt River to the west of the Kazan between 6000 and 8000 years ago.

Distinctive stone tools made by Palaeoindians are found throughout North America, in some places associated with now extinct species of elephant and bison, and also with caribou. Beautifully made, leaf-shaped spear heads of this tradition have been found in Wyoming coulees as well as on barrenland eskers. One of our objectives was to look specifically for Palaeoindian sites along raised beach ridges that date to the early post-glacial period. The search would help establish the limit of the Palaeoindian world in Canada. Was the Kazan valley, like the neighbouring Dubawnt River, occupied by Palaeoindians? Or did the ice linger in the lower Kazan valley for too long, making this region inhospitable for the earliest northern caribou hunters? No Palaeoindian artifacts that would indicate a very early occupation of the area were found.

About 4000 years ago, the first Eskimo people arrived. Palaeoeskimos, as they are known to archaeologists, were the first people to colonize the Arctic, from Siberia to Greenland, on a year-round basis. They eventually moved south into the barrenlands from the arctic coast. This movement coincided with a shift towards colder and drier weather in the north, a probable southward displacement of the treeline, and the disappearance of the material culture associated with Indian settlement in the southern tundra. The Palaeoeskimo technology was based on blades of stone removed in a standard fashion from small cores of chert (flint). These blades were then trimmed and shaped to make components of tools for hunting and domestic activities. They are testimony to an ingenious technology which introduced, among other things, the bow and arrow to North America.

The appearance of ancestors of the modern Chipewyan Indians on the barrenlands is marked by the spread of the Taltheilei tool-making tradition about 2500 years ago. The most recent period of this tradition, the Late Taltheilei, extends until contact was made between the Dene (the people), including the Chipewyan, and Europeans in the 18th century. The Chipewyan, who lived along the treeline in the southern Keewatin, are the most northeastern Dene group. Their way of life, was, like the Caribou Inuit of the more recent post-contact period, based largely on the hunting of caribou. The herds migrated across the treeline twice a year, in spring and fall, when the Chipewyan intercepted them. During the summer, people ventured widely onto the tundra to hunt caribou at water-crossings. It is these water-crossing sites, found commonly along the Thelon and Dubawnt rivers to the west, that we expected to see also on the Kazan.

Both the Palaeoeskimo and the Taltheilei, or early Athapascan Indian people, represent an important part of the story of human settlement of the tundra. Yet little is known about the early history of the Kazan valley. Who were the first people to inhabit this land? Did both these traditions gain a foothold here? How did treeline movement along the Kazan and other north-south river valleys affect the arrival, survival and departure of these cultures?

Despite the succession of different cultural traditions, the fundamental adaptation involving caribou did not change. Like the Chipewyan, earlier peoples probably entered into a close, year-round dependence on the caribou, orienting their seasonal movements around the treeline, north to the tundra in summer and south to the forest in winter.

However plentiful, the caribou were a seasonal resource. To the inhabitants of the northern barrens, they were usually available only during the brief arctic summer. Along the southern edge, the Chipewyan people were able to play out their encounter with the herds for a longer period during the year as the animals migrated through this area seasonally. The critical time and event everywhere along the Kazan was the late summer or early fall hunt. Then, the meat is fatty and most nutritious, the hides are sleek and suitable for making warm clothing. The river itself was (and still is) a barrier to movement, causing the animals to cross at narrows. These water-crossings, in the context of the migratory behaviour of caribou, provided an element of stability or predictability for hunters in an uncertain land.

The massing of animals and the precise route and timing of their migrations are not entirely the same year after year. Weather, wolves, insects, population levels—all have complex repercussions for caribou behaviour. This annual variability had tragic consequences for the Caribou Inuit. Greater numbers of caribou in the pre-contact past may have meant a more reliable annual staging of migration, but there is no definitive confirmation from the archaeological evidence.

The Caribou Inuit penetrated the Kazan region during the 19th century at a time when the Chipewyan Indians, due to both the fur trade and disease, were withdrawing into the forest. The Inuit were the last people to come to the valley and, together with the Chipewyans, they are the only people actually recorded there by early explorers, traders and priests.

The hot afternoon sun glanced off the surface of the water in the broad section of the river where we were paddling. I looked through my binoculars to the west at the long thin shadow of an ancient beach ridge on a hillside. Tyrrell described the same ridge in 1894. Water, dammed by the glacial ice in Yathkyed Lake, etched this line some 7000 years ago. The water may have found a temporary outlet to the south, up what is now a small tributary of the Kazan to another river valley. Or the ice in Yathkyed Lake may have disintegrated quickly, releasing the water through the rest of the Kazan system to the Tyrrell Sea. If so, the earliest visitors to this part of the Kazan saw water

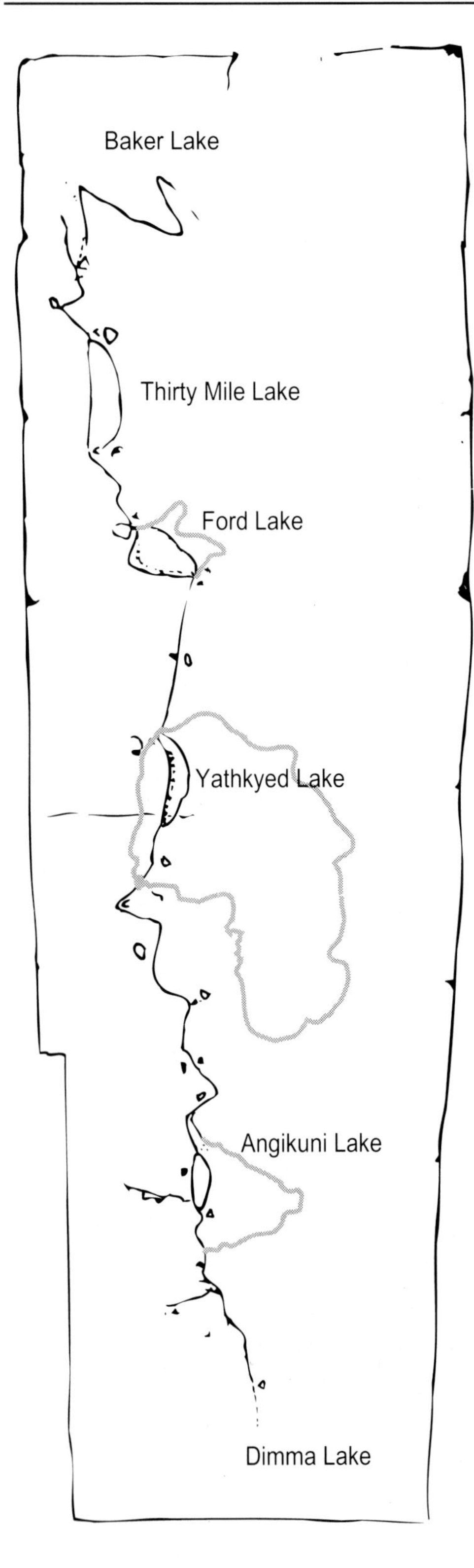

HALLO'S WORLD

Map of the Kazan River drawn by Hallo for J.B. Tyrrell, August, 1894. Hallo's rendering from memory, executed with unfamiliar tools, reflects Inuit use of the river and is remarkably accurate. The shaded areas represent the approximate size and shape of the lakes today, which would be different from the part of the lakes that Hallo used in his travels.

Drawn from a photograph of the original, in the J.B. Tyrrell collection, Thomas Fisher Rare Book Library, University of Toronto.

levels as they are today, and probably camped on approximately the same shoreline as all later peoples in the valley.

We drifted by a point of land, an esker of sand meandering into the water and reflecting white sunlight beneath our canoes. The exposed sand had a promising look — the past revealed in an eroding landscape. We landed on the downstream side to survey the esker. A rapid walk across the eroded floor and around remnant islands of the former land surface was enough to reveal a major pre-contact occupation: the ground was littered with quartzite flakes.

The occasional bifacial stone tool (showing the scars of knapping, or flaking, on both sides) might allow us to identify the culture to which the artifacts belonged. There was a good possibility that material from different peoples using different stone-working traditions had been left at various times in the past on the floor of the eroded esker, together with cobbles too big to have been removed by the action of wind and water over centuries. This suspicion was reinforced when we examined the artifacts more closely.

Large, skillfully flaked, quartzite bifaces suggested the middle Taltheilei tradition, which has been radiocarbon dated to between A.D. 150 and 600. This tradition suggests that ancestors of the Dene camped on the river here about 1500 years ago. Many of the flakes where we were now standing were exotic cherts rather than local quartzites. Cherts occur naturally in the bedrock hundreds of kilometres to the northeast of here. Pieces may have been brought by Palaeoeskimos, who preferred the fine quality of this rock to produce their distinctive blades and small tools. Fortunately, not all of this site was eroded or deflated, with artifacts from presumed different peoples and periods mixed together. Part of the site remained embedded in its original stratigraphic context within a large column of sand, capped and protected by a tuft of gnarled spruce. Future work at this site will be needed to probe the levels representing encampments during different periods in order to establish its antiquity and cultural origins.

Earlier in the day, another group recorded a site that was spilling out the side of a huge sand embankment. Among the artifacts was one chert blade - a hint of the possibility that Palaeoeskimos lived here 3000 to 4000 years ago. We left this region aware of its potential as one of the oldest and culturally most diverse sections of the river.

Paddling past eskers forming soft sandy banks and islands, our canoes entered a narrower, faster current as we approached Yathkyed Lake. The strong, cold breeze off the lake dispelled the sultry atmosphere of the past few days on the river. Standing waves rising against the wind splashed into our canoes. The river curved around a grassy highland that rose almost a hundred metres above the water, commanding a view of the protected valley to the south, and northwards over Tikerarjuak, the "long forefinger" or headland which projects into the vast lake beyond. It was near here that Samuel Hearne encountered a party of Chipewyan families hunting caribou in a river "which empties itself into a large lake called Yath-kyed-whoie, or White Snow Lake."[14]

There has long been the suggestion that the Kazan River was an area of overlap between the Chipewyan Indian and Caribou Inuit. In 1868, Father Gasté, the Oblate missionary, described his journey with the Chipewyan through their summer hunting grounds on the barrens:

> *"It had been mentioned that we would form a single camp at the end of the hunting season, and I had been highly pleased at that decision. When the time came, a few of our Indians, in their haste to see the Eskimos again and to trade with them, prevailed upon the others to leave as quickly as possible and to spend with the Eskimos the time set aside to tarry on that particular spot."*[15]

When Tyrrell paddled around this bend more than a hundred years after Hearne, the cultural landscape had changed. Between Ennadai and the portage north of Yathkyed Lake, he encountered about twenty-five camps of Caribou Inuit hunters and their families, each containing an average of ten to twenty people with as many as 65 at one large settlement. When they first saw him at Ennadai Lake, the Inuit ran away from the white men and their Chipewyan guides. Tyrrell took on Inuit guides to replace the Chipewyan ones who left him at the "northern confines of their hunting grounds" because "the great treeless wastes to the north were supposed to be thickly peopled with unfriendly Eskimos, who would almost certainly destroy them".[16] This, at least, was Tyrrell's impression.

Yet relations between these two peoples were not universally bad. Other accounts, like Gasté's, hint at a rich social fabric and cultural interaction about which we know very little. A high potential for trade existed in the southern Keewatin before and during contact with Europeans. Trade may have included items such as waterproof seal skins, walrus sinew lines, soapstone, and dogs (all from the northwest Hudson Bay coast and adjacent tundra), which may have been exchanged for wood from forested areas to the south. During the 18th century, European goods may have been traded northwards by

"We could see an inuksuk, so we spread out and started to fook for features. We started to realize how extensive this site was.We spent five days there and walked over that site again and again. We kept finding new features. What had been a landscape when we first walked on it, now became a working environment for the people who were once there. Sometimes you would sit on a high spot and try to imagine all those people walking around and working, and putting meat into meat caches and smashing the bone to extract marrow. The landscape became something more than just a beautiful spot for paddlers. It became somebody's home and their working environment - you learn to respect it more."

Hilli Woodward - England

Chipewyans who had much better access to Hudson's Bay Company traders at Churchill. Historic evidence suggests that the southern Keewatin was an active boundary area, extensively used by both Chipewyan and Inuit peoples.[17]

The existence of Yathkyed Lake in two cultural worlds is reflected in its names. Yathkyed is called Hikuligjuaq, or "Great Ice-Filled One" by the Caribou Inuit. From the high hill to the west, we had our first fog-shrouded glimpse of the lake. On July 20 it indeed appeared an enormous, forbidding basin of ice. Waiting for the lake to clear, we made an intensive search of a large region on either side of the river for archaeological sites - both on the wind-blown highlands and in the black fly-infested lowlands, popularly known within the group that did most of the surveying there as "willow hell".

During this search, we found sites a long way inland from the river. They contain similar features. Tent rings, meat caches and inuksuit appear to string out from the main concentration of activity beside the river, as if following or anticipating the movements of caribou to or from the interior. One of these sites is somewhat different. It is the farthest inland of all the sites we found, and isolated from the rest by several kilometres. It is also devoid of meat caches, despite the evidence of tent rings and hunting blinds for extensive use and settlement.

The absence of caches suggests that the inhabitants were not storing game for long term use over the winter. The combination of its location, far from the other sites yet obviously oriented to the same resources, and the evidence for short-term summer occupation sets this interior site apart from its neighbours. Chipewyan hunters who returned to the forest for the winter would be less likely to store meat than the Inuit. Could this site represent a Chipewyan camp, located at a cautious distance from the river and main hunting area where the Inuit were more likely to camp?

The site at Padlerjuaq, which Tyrrell and Hearne both noted in their journals, is an immense area of former activity. There are bits and pieces everywhere that attest to the popularity of this place. Tent rings and meat caches and caribou bone are scattered in clumps along several kilometres of shoreline. An impressive inuksuk, finely balanced by tiny pebbles wedged at its base, gives the site a skyline. The land and the water have reclaimed much: features lie hidden under willow and dwarf birch and submerged in the river.

"Buggy, boggy, soggy, windless, humid . . ." someone wrote anonymously, introducing the comments section of the archaeological site form for Padlerjuaq. We set out to record these surface features just like the previous sites on the river. This one, however, was vast and required days of often frustrating labour. The seemingly simple task of mapping the relative position of features was turned into an ordeal by bugs, rain, and the sheer scale of the site.

Despite the aggravation, a map of the site began to emerge on sodden paper. More than a hundred features occur on a low peninsula. Beginning at the south tip of the peninsula, which is buffered from the oncoming current by a tangle of half-drowned willows, a few tent rings and bone concentrations lead up to a massive inuksuk. Beyond, piled boulder graves and caches are concentrated on slightly higher, though equally damp ground and spill out vaguely over the landscape. Mapping these features in relation to the surrounding area was not easy: the landscape here lacks definition - the land itself is difficult to distinguish from water.

The controlled excavation of a tent ring opened a small window into the past life of some of its inhabitants. The ring enclosed a large space, five or six metres in diameter. The conical tent which stood above it was probably held up with spruce poles, five to ten 10 metres long, too valuable to have been left behind. The poles were covered with caribou hides sewn together. The Inuit tupik, like the Chipewyan tents, may have required the skins of as many as seventy individual caribou - mature bulls killed in August or September after the holes caused by the warble fly larvae had healed.

The stone "walls" of the ring, in fact boulder weights placed around the bottom of the tent, were well-defined except on the south side where a small break appeared, perhaps defining the entrance. To one side of the door, just inside, a hearth may have been used in bad weather, or perhaps maintained as a smudge pit to smoke out insects during the worst of the bug season.

Gently peeling back the sod, we found the refuse of domestic life littering the black, peat floor of the tupik: the tip of an iron file for striking flakes of quartzite to produce sparks; a piece of metal, flattened and shaped into a spear point with a hole through the base for riveting to a bone fore-shaft; tiny coloured glass beads lost from a string pendant. Three artifacts, though, were particularly intriguing.

A piece of birch bark cut from a large tree far to the south, rolled up and stashed away. A piece of tinder? Or perhaps its rolled condition conceals some sign of workmanship, like stitch marks for making a container. These clues may only be revealed when the bark is opened carefully in the conservation laboratory.

A fragment of mica - a transparent film-like rock, with a hole drilled through it - may be part of an ornament. It was an unusual find, perhaps obtained through trade with the Indians.

A single strip of copper, less than a centimetre long, poses another mystery until, ten days later on the wind-swept northern shore of Yathkyed Lake we find an odd assortment of small mechanical parts, some reminiscent of clock components. I think of accounts I have read of the grave of the famous

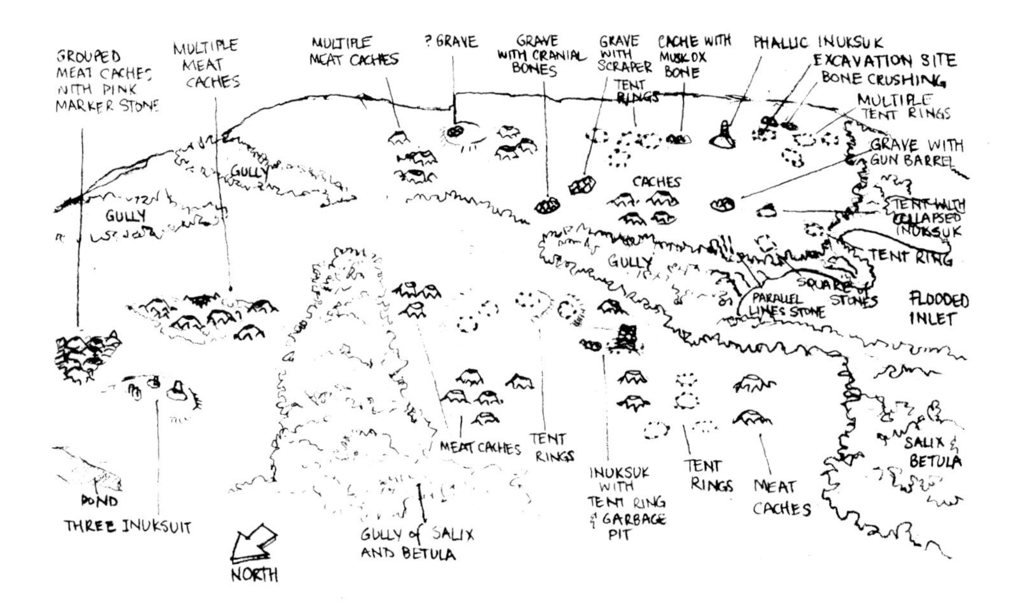

"Like most, I have found the archaeology to be different from what I expected. Learning about sites and site mapping to seeing examples are all one thing for the mind to handle, but when one is trundling about the tundra, one has to look beyond the rocks upon the ground and the bits of wood about and search for those patterns that may exist, or that bit of nature unnaturally placed. It's quite a lot like what I imagine detective work to be (or indeed my own research) little clues here and there that one must first recognize as clues, then step back from to construct the patterns that link them all. First there was Unguluk's camp and partially or mostly sunken stones upon a rise overlooking the river forming three rough tent rings, a few hearths, and possibly a drying rack for a kayak. At least that's what it "looked" like. Heading down the river a bit we find a caribou crossing with loads of bones where the hunters may have slaughtered the animals. Finally, at the end of the transect we find an enormous stretch of ice-cracked rock, littered with the bones of a hundred meat caches and at least one grave near some old hunting blinds. Suddenly, the odd bit of bone or the shimmed up rock take on new meaning when seen in the context of those who lived in this land before us. It's an incredible feeling, to look upon the land and the features man has imposed upon it in much the same manner as they looked out hundreds of years ago,for this is the same land!

Scott McNamee - United States
July 10

hunter Kakoot, where a Victrola and an alarm clock are among a plethora of respectful offerings. The site on Yathkyed Lake where we stood was no grave. Yet clues of its former inhabitants had an other-worldly quality. The little copper strip turned up again - this time a pair of them attached to a piece of metal which we could interpret at last as a harmonica reed.

Finally, over a thousand discarded bone fragments were recovered from inside this tent ring, representing the remains of at least four caribou and two lake trout. A high density of bone was found around the inside periphery of the south half of the ring. This pattern reinforced our impression that this half contained the entrance and was the cooking or food-preparation area. The north half, with a sparser scatter of bones, appeared to have been the sleeping area.

Most of the identifiable caribou bone fragments which were excavated represent the middle of the animals: ribs and vertebrae in particular. The relative absence of bones from the animals' heads (skull, jaw, teeth, and antler) recalls the special treatment sometimes accorded to this part of the caribou. Knud Rasmussen, the Danish ethnographer who visited the Paatlirmiut on the north side of Yathkyed Lake in 1923, describes a feast given in his honour by Igjugarjuk, a leading hunter and shaman, indirectly suggesting a reason why these bones may be missing from the refuse in the tent ring at Padlerjuaq:

> *"The two cooked caribou lay in great carved joints over the floor, divided in wooden trays, and as Igjugarjuk had wisely thought that our habits might not be quite the same as theirs, he had placed our rations on a separate dish. . . Besides the two caribou a number of heads had been cooked. Each member of the expedition received one . . . Only this special condition was attached to the gift of meat, that each one of us had to guarantee that the leaving would not be gnawed by women or dogs. The muzzles especially were regarded as being sacred eating, and this must not be violated."*[18]

This passage suggests that crania were disposed of carefully rather than casually - perhaps at the edge of the site, under rocks, or in the river. This is not the only explanation for the absence of cranial bones in the tent ring. The carcasses may have been shared among several hunters and their families, the prized heads going to the hunter or hunters who actually killed the animals. If so, the tent ring we excavated may have been lived in by a hunter who helped in the kills. The predominance of hind over front limb bones in the tent ring may support this scenario: Birket-Smith, the leading ethnographer of the Caribou Inuit, observed in 1924 that helpers tend to receive the hindparts of the animal while the foreparts go to the main hunter.[19]

A further alternative is that the heads were consumed at the kill or butchery site, and never made it back into camp. A snack of tongue and other cranial delicacies may have been irresistable to the hungry hunters, especially if the kill took place some distance from the settlement. Each of these explanations would require further work to confirm or deny, involving excavation of bone assemblages in places where killing, butchering and disposal occurred.

Before removal, the locations of all finds were recorded - both horizontally within the living space defined by the walls of the tent ring, and vertically in relation to the surface of the ground. Discoloured soil or features below the ground surface may represent modification of the tent floor while it was inhabited - scorch marks from the hearth for instance, or removal of vegetation for sleeping. Features that at first appear to be cultural in origin may turn out to be natural disturbances, often the result of frost heave. These anomalies, as much as the artifacts themselves, are part of the history of the site.

"We have lost sight of the fact that the features we are walking over were actually constructed by people that lived on this same land a thousand years ago. All the work we are doing seems fascinating. I want to map sites forever, but standing in the middle of a willow swamp, on a pile of arbitrarily arranged stones, taking a bearing to some point in the distance, with the bugs having a bite-a-thon on your face, or the pelting rain making you soggy and fogging your compass - then archaeology loses some of its glamour and one loses sight of the significance of one's work. It is at times like this that it is hard to imagine a pile of rocks with bones as anything more than just that. The imagination that archaeology requires - to bring the ancient peoples back to life and make you yearn to learn more of them - the imagination that is required for all these things - has great difficulty working under such adverse conditions.

"I'm glad we had a chance to do an excavation. It gives us a more varied and complex view of archaeology on the barrenlands. I have tried to follow its progression throughout, for personal interest and to help me get a fair overview of the whole process. The original marking out of the tent ring, the dividing it into quadrants, the mapping of the whole excavation site - to the detail of positioning each stone on a map - the peeling away of the greenery, then the mossy layer, etc. and so on down until you reach the sand. All sorts of odd

bits have appeared in the site -brass buttons, coloured beads, bones, flints, glass, a sawn-off bullet, a metal rivet, and who knows what awaits our probing fingers and trowels. This has been a wonderful site: inuksuit, meat caches, wooden artifacts, tin cans, working with another group. All these things I have enjoyed and the fact that I go on learning - always."

Sonia Mellor - Australia
July 23

In the end, Padlerjuaq may lack dramatic evidence for Chipewyan-Inuit aggregations near Yathkyed Lake. Perhaps these gatherings took place only sporadically when local conditions allowed people to live together in one place and conduct feasts. Or perhaps the important aspect of meeting between peoples from north and south was simply the exchange of goods unavailable locally to each group. This exchange may have taken place at ordinary summer Inuit encampments such as the ones Tyrrell witnessed. Trade between individuals from different camps may not be visible to archaeologists who are looking for evidence of large gatherings in special structures, large tents, and patterned arrangements of features. Painstaking analysis of individual tent rings will be necessary to recover exotic clues, like birch bark and mica, among the discarded or forgotten artifacts of everyday life.

An enormous hill of bald rock with a patchwork of peat and grass loomed before us. From our vantage at its base, an unnatural crown of boulders was the only sign of former human occupation. Two days later, we had discovered and mapped about 70 features which transformed the mountain, in our mind's eye, from a lonely wilderness to an imagined "city" of activity. In fact, on the summit, the clear, low light of those unusually warm days brought to life the monoliths and simple rock structures that seemed on occasion to move when seen indirectly. The height was modest, perhaps only thirty metres above the sparkling river - out of all proportion to the spectacular view of hundreds of lakes and ponds on the soft tundra below.

Tyrrell stopped at an Inuit camp here a hundred years ago. Although he described it as a large camp then, the size of the site which we mapped is probably considerably bigger, reflecting many years of seasonal hunting and fishing. Different kinds of features indicate a wide range of activities. We looked for order in the arrangement of the features on our new site map:

hearths and tent rings on top of the rock where constant exposure to the wind provided relief from insects in summer; mysterious boulder constructions also on top - perhaps drying racks for meat and hides taken in late summer; pits excavated into a rubble substrate on the slopes which contained kayak parts, snow knives for making snow houses, and other artifacts suggesting a mixture of summer and winter use; and graves around the base of the hill marked with long wooden poles. On the ground, the evidence of activity seems randomly distributed over the mountain - the overall arrangement of features is difficult to grasp around the curved slope. The map, however, suggests defined areas of use - for living, working and storage, and for the dead.

> *"The work was often tedious. We knew the menace of bugs on windless days, and the problems of working in adverse weather. But a story unfolded, as more was found and mapped, and the barrenlands landscape became the working environment of a people who had learned its rhythm."*
>
> *Kassie Heath - Australia*

One of the principal goals of the expedition was to record as many different types of sites and features in as wide a range of environmental settings as possible. The Caribou Inuit lived in the interior year round, hunting, fishing, and trapping, based in small overnight camps, as well as in more permanent seasonal settlements lasting a few weeks to a couple of months. Our work ultimately was to map the Kazan valley, the distribution of sites in relation to important aspects of their environment: fishing eddies and rapids, caribou crossings, and sources of stone. With enough sites recorded, patterns of settlement along the Kazan may emerge, and these can be compared to patterns in other parts of the barrenlands. Given the unique way of life of the Caribou Inuit, the patterns were not expected to be quite like anything known.

Caribou hunting certainly played a prominent role in the history of the Kazan River, and so we expected to find many large sites at places where caribou could be intercepted and killed during their annual migrations. For instance, the thousands of caribou that streamed past us near Yathkyed Lake went right through a large site where, for years, hunters and their families established camp during the spring to intercept the animals' northward approach. During our approach to the site from the lake, what first drew our attention was a line of dark rocks against the sky, perched unnaturally on bedrock ridges. Islands of bedrock, emerging from the tundra like the swell from a sea of glacial till, commanded a view of Yathkyed Lake on one side, and a soft, damp landscape everywhere else. At our arrival, the low-lying land was swept by a rising tide of moving fur and antlers. From below, we saw the silhouettes of a classic inuksuit formation that once helped to channel migrating caribou towards hunters hiding behind rock blinds. Scattered across the ridge was the cumulative evidence of years of meals. Spirally-fractured fragments of bone are testimony to the practice of breaking leg bones, in particular, to extract the rich marrow from inside.

Farther north, on another ridge, another line of inuksuit leads down to the water's edge. This is a place near the outlet of the lake where the caribou have been observed to cross the river for generations. Topped with peat, sprouting grass, and sometimes with the wings of gulls or scapulae that flapped crazily in the wind, the upright stones once took on a more lively appearance. Women and children would drive the caribou towards this threatening but illusory inuksuit barrier. The animals, rather than crossing the line, took to the water where they were ambushed by hunters in kayaks armed with long lances.

This kind of site encompasses all the activities of a major successful hunt: the preparation and anticipation; the ambush; the butchering of carcasses, drying of meat and hides; and sometimes storage for later consumption. Over years, these activities are repeated in somewhat different locations at the same river-crossing and eventually overlap and merge with one another creating vast sites. The usual surviving remnants, stone tools and flakes discarded during the knapping process of stone tool manufacture, represent only the tip of the iceberg, for the organic component of the record - especially the vast quantities of bone as the caribou were carved up and consumed - are dispersed by scavengers and disintegrate quickly in the acidic conditions of the northern environment. This activity was repeated at hundreds of crossings along the Kazan. The pattern may extend through time as well as space. At places where sediment is likely to bury evidence of such behaviour, such as at the eskers we passed during the first part of our trip, assemblages of tools may survive for thousands of years.

This pattern of activity and settlement, found throughout the barrenlands during the pre-contact period, attests to the importance of caribou to all native cultures regardless of ethnic origin. The people of the Kazan are no exception, indeed they are probably the most dramatic example of this dependence. To compensate for the uncertainties of caribou movement from year to year, people sometimes travelled great distances. The Chipewyan seasonally moved from the forest on to the tundra to intercept the caribou returning from the calving grounds. The Inuit who stayed on the barrens year round travelled vast distances between reliable hunting and fishing locations and to recover caches of meat left at river crossings after the fall hunt. [20]

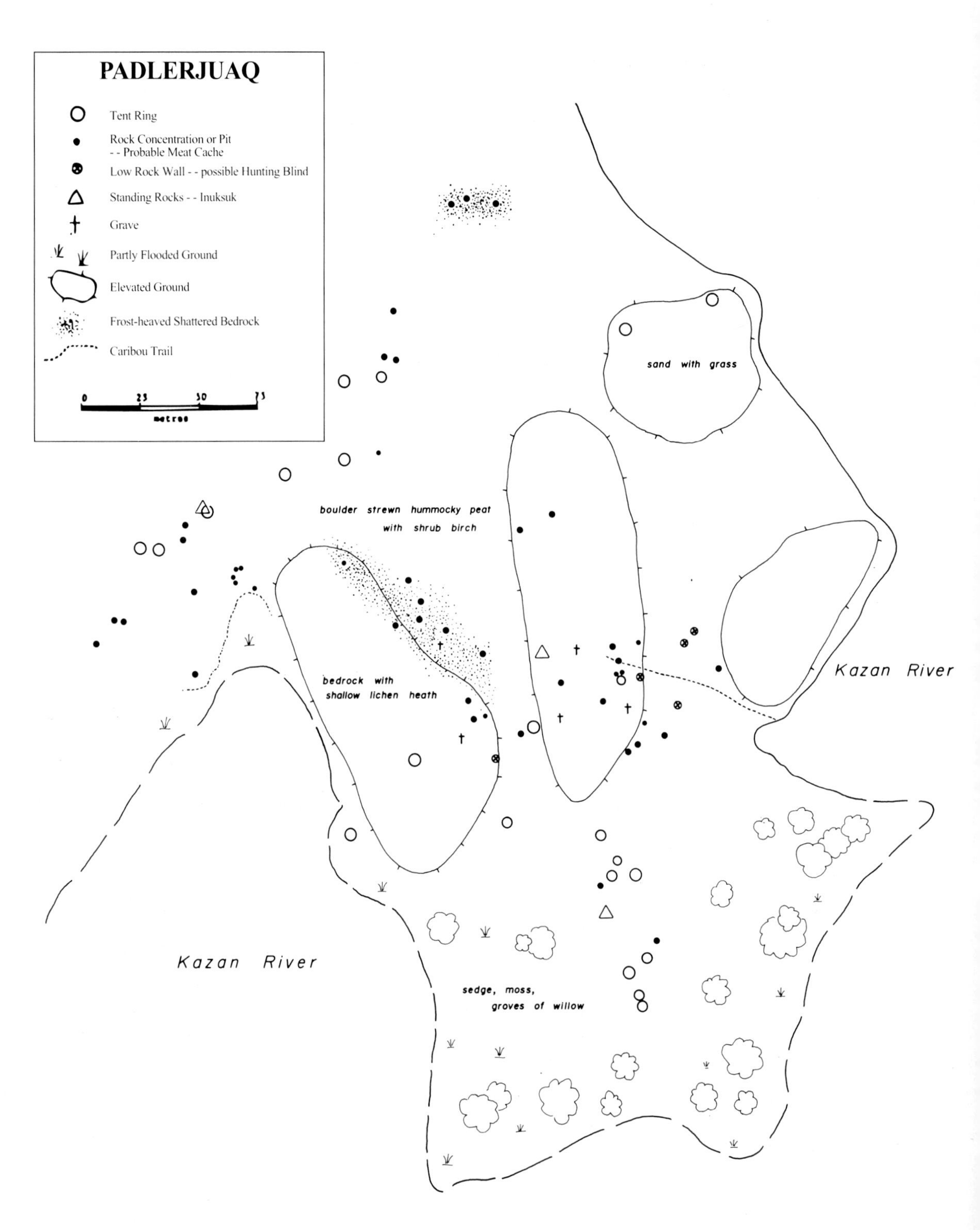
PADLERJUAQ
Tent Ring
Rock Concentration or Pit
- - Probable Meat Cache
Low Rock Wall - - possible Hunting Blind
Standing Rocks - - Inuksuk
Grave
Partly Flooded Ground
Elevated Ground
Frost-heaved Shattered Bedrock
Caribou Trail
0
25
50
75
metres
sand with grass
boulder strewn hummocky peat
with shrub birch
bedrock with
shallow lichen heath
Kazan River
Kazan River
sedge, moss,
groves of willow

An economy of effort, however, was maintained. Caribou were not chased but strategically intercepted during their migrations. Food was pursued on the trail rather than being toted about. Equipment was pulled by dogs whenever possible and much of the overland travel was done by qamutik (sled) during the eight or nine months of the year when ice or snow made this kind of travel possible.The embodiment of this economy was the caribou, which provided sustenance and raw material for a life's needs. Frozen fresh and dried meat were the staples. Bone marrow was rendered and stored to provide fat needed to supplement the otherwise lean spring and summer diet of fish. Appetizing delicacies like tongue, fetus, roast hoof, and warble fly larva, were seasonal perks. Feast fare was more complex, foods like the beeatee of the Chipewyan which Samuel Hearne describes with relish, were only made at caribou kill sites when the ingredients were available:

> *"The most remarkable dish . . . is blood mixed with the half-digested food which is found in the deer's stomach or paunch, and boiled up with a sufficient quantity of water, to make it of the consistence of pease-pottage. Some fat and scraps of tender flesh are also shred small and boiled with it. To render this dish more palatable, they have a method of mixing the blood with the contents of the stomach in the paunch itself, and hanging it up in the heat and smoke of the fire for several days; which puts the whole mass into a state of fermentation, and gives it such an agreeable acid taste, that were it not for prejudice, it might be eaten by those who have the nicest palates.'*[21]

Household wares were manufactured from antler, bone, hide and sinew - all available from caribou - as well as locally procured stone, and metal and wood obtained by trade. Together, these raw materials provide the basis for flexible and resilient tools.

The most critical technological adaptation was fashioned from the simplest, most abundant, and perishable of materials - snow. The snow house, or *iglu*, made it possible to survive in the interior during the winter. Its adoption by the coastal Inuit in the early 19th century from the Igloolik Inuit further north promoted the interior expansion of the Caribou Inuit. It replaced the stone house, chinked with mud and vegetation, which was used formerly on the coast but was impractical for the interior as it tied people down to one location for the winter. The snow house, on the other hand, could be built quickly where it was needed, such as directly on top of a lake where fish and fresh water could be procured.

The snow house illustrates the importance of knowledge and experience over technology among the Inuit. Ingenuity was brought to bear with limited raw materials. Analysis of snow conditions was essential to the successful construction of a snow house. Inuit terminology still reflects this knowledge. New, loose snow is distinguished from firm snow, the best for winter construction, and from hard-packed drift snow, used for spring construction when the rising sun might weaken softer snow. Long thin sticks are used to probe the snow to assess its condition. Special snow knives, which the Hudson's Bay Company eventually came to manufacture for trade, were used to cut the building blocks. The technology was simple, the knowledge required to use it effectively, very sophisticated.

We explored the soft shores of a tributary river, the mud already crowded with the hoof imprints of a herd of muskoxen. Walking upstream on one side of the tributary river, through tall grasses and wet sedges, we came upon a massive ring of boulders encircling a shallow depression or pit. On one side, bushy thickets of willow filtered a view of rapids and rocky pools in the river. On the other, a lush sedge meadow of brilliant green faded into the distance where a line of hills rimmed the northwest horizon.

A spidery network of angular, sharp boulders spread over the surface of the ground pushed upwards by frost over thousands of years. At a widening of this stony course, people at some time in the past had created a floor, slightly below ground level, an internal circular area five metres in diameter surrounded by a ring of boulders thrown out from the cleared centre. This structure, and similar ones nearby, were unlike any others we had seen along the Kazan. The massive construction and the semi-subterranean interior resembled, more than anything else, the houses used by Thule people during fall and winter along the coast of Hudson Bay hundreds of years ago.

The interior was filled with a profusion of willow. The shelter of the ring and the accumulation of organic domestic waste inside the dwelling encouraged this growth. Inside the pit was a pile of rocks almost two metres across with a central depression. It was as if some of the boulders during the original construction of the ring had been reshaped into an enormous hearth. More likely, the walls of the dwelling had been cannibalized to make a meat cache or fox trap, now collapsed, sometime after the house had been abandoned for the last time.

Signs of recent occupation, like the wooden tent pole fragment wedged into the top of the rock ring sometime within the last century, contrast with the ancient appearance of the ring itself. Did Thule people overwinter here on the Kazan centuries before the Caribou Inuit? Did the Inuit sometimes build, in the tradition of their remote Thule ancestors, winter dwellings of stone and sod with roofs of hide? Perhaps this place was a kind of oasis where muskoxen, caribou and fish could be reliably caught and cached, inspiring substantial houses to be built in the autumn and occupied through the winter in lieu of the more usual snowhouse. The answers to these questions lie under the willow and turf, but the broad outlines of the story are clear: a complex history of construction, use, reuse, and modification for new purposes. A familiar story on the Kazan is repeated. Old ways were adapted to new demands. They survive, in one form or another, alongside more efficient techniques.

> *"I see the inuksuit today and feel their age. I see in my mind solitary people using them maybe five times in a generation, and it creates a real sense of companionship with those who have gone before. I am separated by the greatest distance of all – time. Yet I am seeing basically the same land. Distances of place, time, and experience."*
>
> *Scott McNamee - United States*

The archaeological record of the Kazan embodies at least two remarkable links. One is symbolized by the possibility of cultural exchange between Indian and Inuit in this region. The other is between the recent and unique way of life of the Caribou Inuit, which included hunting and trapping on the tundra year-round until the 1950s, and an ancient respect for the land.

The expedition sought to derive an understanding of the significance of the Kazan valley within the broader pattern of human occupation of the barrenlands. Although it has long been recognized that Chipewyan and Caribou Inuit hunters utilized the area in the period after European contact, no systematic archaeological examination of the region had ever been undertaken. Therefore, little was known about the antiquity, sequence, or extent of the pre-contact occupation. No substantial attempt had ever been made to examine the connection between the Inuit, who were the last residents of the river, and the earlier occupants. The expedition attempted to document the links between the traditional cultures, and looked for clues to the arrival of people in the remote past, through the recording and interpretation of over 200 sites. Taken as a whole, these sites establish patterns of land use representing both long-term human occupation and yearly, seasonal use by different groups in the Kazan valley.

The larger message that emerges from this evidence is one of traditional peoples' reliance on the land. Their camps speak of times of plenty and their graves evoke the desolation of disaster. Life depended upon intercepting the caribou, not rolling your kayak as you hunted at a river crossing, knowing where to find muskox, and driving caribou successfully to an ambush. To the people, the Kazan was a river of life, sustaining cultures whose traditions were drawn from a deep and ancient knowledge of the land, as well as from life experience on the shores of Angikuni, Hikuligjuaq, and the Kazan River. The remains of these cultures everywhere in the valley bear silent witness to the hunters' life in an emerging land.

Andrew Stewart - Canada
Expedition Archaeologist

Further Reading

Birket-Smith, K. 1929. "The Caribou Eskimos; Material and Social Life and Their Cultural Position" in *Report of the Fifth Thule Expedition* 1921 - 24. Vol. 5, Nos 1-2. Nordisk Forlag,Copenhagen.

Burch, E., Jr. 1986. "The Caribou Inuit" in *Native Peoples, the Canadian Experience*, edited by R.B. Morrison and C.R. Wilson, pp106-133. McClelland and Stewart, Toronto.

Damas, D., (ed.) 1984. *Arctic. Handbook of North American Indians*,edited by W. Sturtevant, Vol. 5. Smithsonian Institution,Washington.

Helm, J. (ed.) 1981. *Subarctic. Handbook of North American Indians*,edited by W. Sturtevant, Vol. 6. Smithsonian Institution,Washington.

McGhee, R. 1978. *Canadian Arctic Prehistory*. Van Nostrand ReinholdLtd., Toronto.

Smith, J. and E. Burch, Jr. "Chipewyan and Inuit in the Central Canadian Subarctic" *Arctic Anthropology* 16 (1979): 76-101

"It was easy to think, before I arrived in Canada, that this was going to be a real exploring adventure. In many ways, I still believe this to be the case, but I find I am constantly reminded that this is a journey into a land with a rich and colourful history, a land that is wild, wonderful, and windswept. It is a land that emerged thousands of years ago from under the powerful ice sheet. It is a land on which man has not only survived, but thrived. It is an emerging land in the sense that we're beginning to discover its beauty, both physically and in terms of cultural heritage. Now ***we*** *are emerging from our ignorance of the North, rather than just the land, which rises magically above my wildest imaginations. It is a land that inspires."*

Richard Wilson - Scotland

JOURNEY INTO AN EMERGING LAND

Setting out to study a river valley is not unlike the process of studying any other living organism. If one were to understand the human body, it would be essential to first learn about the component parts: the various organs, the skeleton, blood, bone and tissue. In due course, it might become possible to step back, to examine the whole, to see it as a complex system of many parts, and to understand the whole. The ecosystem of the Kazan Valley is not much different.

Although each research project in itself was designed to serve its scientific purpose, it is when they are combined as a whole that the efforts of the expedition achieve their real significance. Each of the scientists behind the projects has inevitably retained the specific focus of their discipline, but each has gained the advantage of seeing their part of the multi-layered system in that larger context.

The land spoke to everyone involved in different ways. For some of the young people it was an awakening to the natural environment. The experience will probably be credited with changing their lives. For each of the scientists, taken individually, the data collected offer new clues to understanding the mysteries that are central to their professional lives. But on a grander, more detached plane, the land spoke to everyone with a message about its state of being. Beyond all the scientific disciplines, with a perspective that defies exact measurement, we begin to see the land for what it is: a multi-layered tapestry of inter-related systems.

The Inuit and Indian hunters who once lived in the Kazan Valley were absolutely dependent upon the caribou. The starvations as recent as the 1950s prove that. Other animals, too, were an important part of the matrix that was their life. The caribou, for their part, are dependent upon the lichen and plant life of the barrenlands. Other animals, like the arctic fox or the snowy owl, rely on the population of lemmings for their survival. The interdependencies are legion.

Life on the tundra is based upon adaptations that maximize the use of available solar energy. Radiant solar energy often overrides temperature, kick-starting life into spring even before winter has let go, by providing direct jolts of heat to plants and animals. The constant interplay of predator and prey allows the secondary transfer of solar energy from vegetation to omnivores and carnivores through a relatively small number of herbivores. In the matrix of life on the barrens, Man is just one element in the system, as dependent on the whole as is the wolf or the lemming, the falcon or the longspur.

Combining an archaeological survey with a breeding bird census, and tree-coring with caribou counting, brings that matrix into focus in a spectacular way.

The idea that the inter-relatedness of the scientific surveys parallels the multi-layered complexities of a wilderness journey was integral to the expedition. A scientist's perspective is formed as much by his mode of travel, as it is by research design. For the scientist in every member of the expedition, the canoe allowed the Kazan Valley to emerge with the same rhythmic association of land, water and people that it offered the first European travellers and the Native people before that. Canoe travel on an arctic lake edges you over the horizon into an ever-new, constantly changing vista. That is discovery beyond the confines of any single scientific discipline.

The intimacy afforded by observation from a canoe, or by walking on the tundra, was important to providing our international group with a sense of the land. It is not possible to be a good observer without that familiarity. Though we knew the land had been inhabited, our experience was an unpeopled wilderness. An intellectual knowledge that Native people lived on the land is not the same as finding their camps, recording the scraps of caribou bone and discarded tools, while caribou from the Kaminuriak herd flow around you. It is in such profound ways that the land speaks. So the valley emerges, and a multi-layered understanding takes root in the minds of today's explorers.

The journey into this emerging land embarked upon by the internationally composed Canadian Arctic Expedition could stand as a turning point. This broad-based, multidisciplinary examination of the Kazan Valley offers a new perspective of this barrenland. If the wilderness is to last, protection alone is insufficient. The land must be honoured and understood.

Every one of our canoeists cast a vote of support for the Kazan River's designation as a Canadian Heritage River. That status was accorded in the summer of 1990, marking the beginning of a new management regime.

The valley has evolved essentially to the extent of nature's power. Given present conditions, the ecosystem has matured. Conditions will change in the future, however, this time in ways more directly caused by Man and not by nature alone. Whereas before, men were no more than players on the stage that was the barrenlands, today we become the stage managers. Man's role changes, but the journey into an emerging land continues.

Christopher C. Hanks - Canada
Expedition Chief Scientist

Mary McCreadie - Canada
Expedition Group Leader

David F. Pelly - Canada
Expedition Leader

APPENDICES

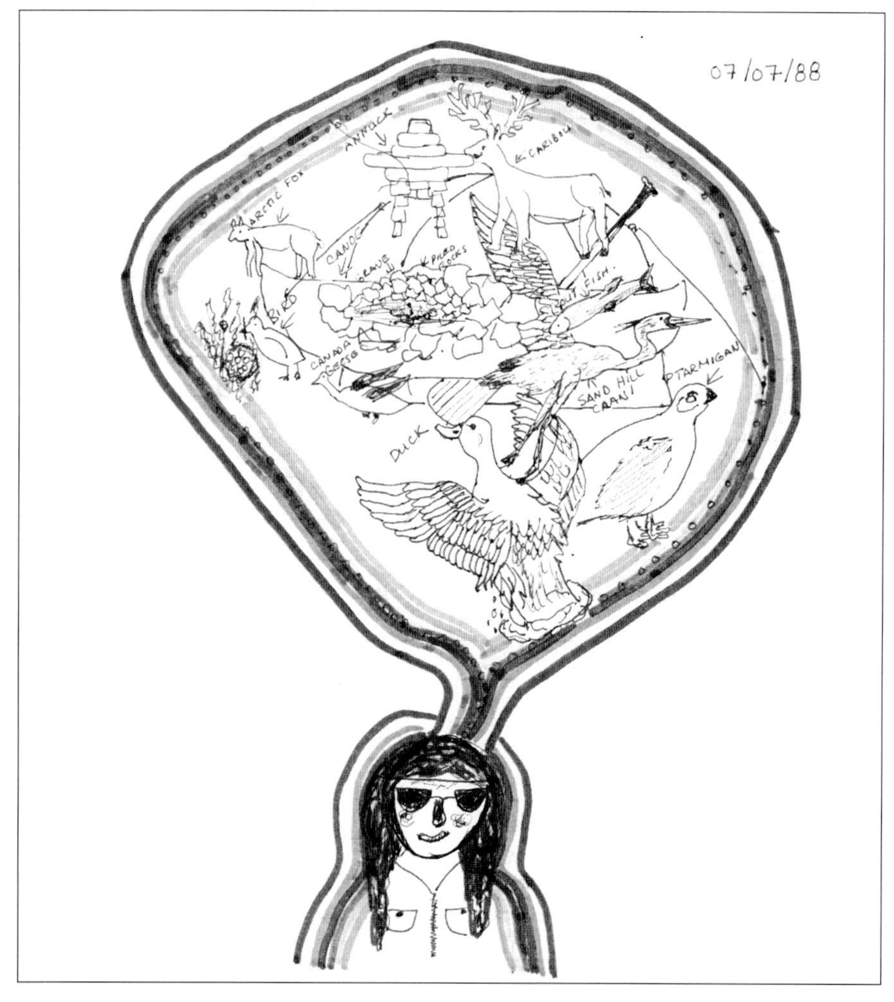

Betty Ann Betsedea
Willowlake River, NWT, Canada

Appendix 1
BREEDING BIRD ATLAS PROJECT

OBSERVED

O Species (male or female) observed in a square during its breeding season, but no evidence of breeding. Not in suitable nesting habitat. Includes wide-ranging species such as vultures or raptors, or a colonial species not at the nesting colony.

POSSIBLE - (PO)

✓ Species (male or female) observed in suitable nesting habitat during its breeding season.

✗ Singing male present in suitable nesting habitat during its breeding season.

PROBABLE - (PR)

P Pair observed in suitable habitat during its breeding season.

S Permanent territory presumed through song at same location on at least two occasions 7 or more days apart.

T Permanent territory presumed through defence of territory (chasing individual(s) of the same species).

C Courtship behaviour or copulation.

N Visiting probable nest-site.

A Agitated behaviour or anxiety calls from adult.

B Excavation of hole by woodpecker.

CONFIRMED - (CO)

CN Carrying nesting material, such as sticks or other material. Please submit full details including location within the square of the observation.

NB Nest building at the actual nest-site.

PE Physiological evidence of breeding (i.e. highly vascularized, edematous incubation [brood] patch, or egg in oviduct. To be used only by experienced bird banders on local birds during the nesting season.)

DD Distraction display or injury feigning.

UN Used nest or eggshell found. Caution: these must be carefully identified if they are to be accepted.

PY Precocial young. Flightless young of precocial species restricted to their natal area by dependence on the adults or limited ability.

FL Recently fledged young (either precocial or altricial) incapable of sustained flight, restricted to natal area by dependence on adults or limited ability.

ON Occupied nest: adult entering or leaving a nest-site in circumstances indicating occupied nest: includes adult sitting on nest. To be used for nests which are, for example, too high for the contents to be seen.

CF Carrying food: adult carrying food for the young.

FY Adult feeding recently fledged young.

FS Adult carrying fecal sac.

NE Nest with egg(s).

NY Nest with young seen or heard.*

• Presence of cowbird eggs or young is confirmation of both the cowbird and the host species.

Appendix 2
SPECIES LIST AND BREEDING STATUS, KAZAN RIVER EXPEDITION

BREEDING STATUS

O	Observed	2
PO	Possible	10
PR	Probable	9
CO	Confirmed	31
		52 species

COMMON NAME	SCIENTIFIC NAME	BREEDING STATUS CODE
Red-throated loon	Gavia stellata	PR
Pacific loon	Gavia pacifica	PO
Common loon	Gavia immer	CO
Yellow-billed Loon	Gavia adamsii	PR
Tundra swan	Cygnus columbianus	CO
Gr. White-fronted goose	Anser albifrons	CO
Snow goose	Chen caerulescens	CO
Canada goose	Branta canadensis	CO
Northern Ppintail	Anas acuta	CO
Greater scaup	Aythya marila	PR
Oldsquaw	Clangula hyemalis	CO
Black scoter	Melanitta nigra	PO
White-winged scoter	Melanitta fusca	O
Common merganser	Mergus merganser	PR
Red-breasted merganser	Mergus serrator	CO
Northern harrier	Circus cyaneus	PO
Rough-legged hawk	Buteo lagopus	CO
Peregrine falcon	Falco peregrinus	CO
Willow ptarmigan	Lagopus lagopus	CO
Rock ptarmigan	Lagopus mutus	CO
Sandhill crane	Grus canadensis	CO
Lesser golden-plover	Pluvialis dominica	CO
Semipalmated plover	Charadrius semipalmatus	CO
Whimbrel	Numenius phaeopus	PR
Semipalmated sandpiper	Calidris pusilla	CO
Least sandpiper	Calidris minutilla	PO
Pectoral sandpiper	Calidris melanotos	PO
Dunlin	Calidris alpina	PO
Stilt sandpiper	Calidris himantopus	PO
Red-necked phalarope	Phalaropus lobatus	CO

COMMON NAME	**SCIENTIFIC NAME**	**BREEDING STATUS CODE**
Parasitic jaeger	Stercorarius parasiticus	PR
Long-tailed jaeger	Stercorarius longicaudus	PR
Herring gull	Larus argentatus	CO
Arctic tern	Sterna paradisea	CO
Snowy owl	Nyctea scandiaca	PO
Short-eared owl	Asio flammeus	PO
Horned lark	Eremophila alpestris	CO
Common raven	Corvus corax	PO
Gray-cheeked thrush	Catharus minimus	PR
American robin	Turdus migratorius	PR
Water pipit	Anthus spinoletta	CO
Blackpoll warbler	Dendroica striata	CO
American tree sparrow	Spizella arborea	CO
Savannah sparrow	Passerculus sandwichensis	CO
White-crowned sparrow	Zonotrichia leucophrys	CO
Harris' sparrow	Zonotrichia querula	CO
Lapland longspur	Calcarius lapponicus	CO
Smith's longspur	Calcarius pictus	CO
Snow bunting	Plectrophenix nivalis	CO
Rusty blackbird	Euphagus carolinus	CO
Common redpoll	Carduelis flammea	CO
Hoary redpoll	Carduelis hornemanni	O

Appendix 3
CANADIAN ARCTIC EXPEDITION

EQUIPMENT

Canoes: 17' Old Town "Trippers". Two people to a boat.

Paddles: laminated paddles with resin-tipped blades by Grey Owl.

PFDs: canoe vests with cargo pockets by Mustang.

Tents: Geom EX 3 & 5, Sentinel, and Traverse, and Denali tents by Eureka!

Sleeping Bags: 3-season Pioneer bags by Jones.

Sleeping Pads: blue airolite pads from Trail Head.

Canoe Packs: grey and red 85 litre canoe packs by Camp Trails.

Personal Packs: convertible fanny/day packs by Camp Trails.

Rescue Ropes: 70' throw bags supplied by Trail Head.

Rescue Z-Drag Equipment: supplied by Canadian Mountain Supplies.

Personal Clothing: full outfits supplied by Tilley Endurables.

Pile Sweaters: supplied by Patagonia.

Insect Repellent: 95% DEET by Deep Woods.

Stoves: Peak I naptha stoves by Coleman.

Ovens: folding ovens by Coleman.

Fireboxes: 2-pot fold down, environmentally sound fireboxes by Trail Head.

Kitchen Packs: supplied by Trail Head.

Cameras: 35mm rangefinders on loan from Konica.

Film: a variety of stock by Konica and Kodak.

Processing: by BGM.

Storm Suits: full Gore Tex Suits by Banff Designs.

Compasses: custom Rangers and standard models by Silva.

Plastic Containers: variety by Nalgene.

Rubber Boots: Hunter Wellingtons by Gates of England.

HF Radios: field units, base-station and radio/telephone interconnect from Communications Canada.

VHF Radios: two channel handsets by ICOM.

Emergency Locator Beacons: by Marconi.

Basecamp Communication: Fax Machine by Sanyo.

Food: supplied by George Weston Limited.

Transportation: VIA Rail, Calm Air, Canadian Airlines International.

SPONSORS

FINANCIAL

Beatrice Canada Ltd.

Beaverbrook Foundation of Canada

Burns Fry Ltd.

Campbell's Soup

Canadian Imperial Bank of Commerce

Canadian National Sportsmen's Shows

Cara Operations Ltd.

Crownx Inc.

Douglas Basset

The Eaton Foundation

Foreshore Project Ltd.

Horn Abbot/Trivial Pursuit

Imperial Oil Limited

Ivex Investments Ltd.

jj Barnicke

Johnson Wax
Nabisco Brands Ltd.
Manufacturers Life Insurance Co.
McDonald's Restaurants of Canada Ltd.
McLean Foundation
Merrill Lynch Canada Ltd.
The Rathgeb Foundation
Ronald A. Chisholm Ltd.
Mr. P.R. Sandwell
Sonar Foundation
Sporting Life
Texaco Canada Inc.
Thorne Ernst and Whinney Ltd.
Toronto Dominion Bank
Toronto Sun
George Weston Ltd.

SCIENTIFIC

Agriculture Canada
Archaeological Services Inc.
Assn. of Canadian Universities for Northern Studies
Canadian Nature Federation
Department of Culture, Government of the NWT
Department of Renewable Resources, Government of the NWT
Indian and Northern Affairs
Keewatin Inuit Association
McMaster University
Prince of Wales Northern Heritage Centre
Royal Canadian Geographical Society
Royal Ontario Museum
Science Institute of the Northwest Territories
Trent University
World Wildlife Fund Canada

EQUIPMENT

Banff Designs
Camp Trails
Canadian Marconi Company
Canadian Mountain Supplies
Coleman Canada
Department of Communications, Government of Canada
Eureka! Tents
Everest Manufacturing
Grey Owl Paddles
Johnson Diversified Inc.
Jones Leisure Products
Konica Canada Inc.
Mustang Industries
Nalgene Trail Products
Patagonia
Sanyo Canada Inc.
Silva
Tilley Endurables
Trail Head Ltd.
Wanapitei Co-ed Camps Ltd.

SERVICES

Baton Broadcasting
Betelgeuse Books
BGM Colour Laboratories Ltd.
Calm Air
Canadian Airlines International
DHL International Ltd.
Edviron Services
The Printing House
Queen's University Faculty of Education
Trail Head Ltd.
Undersea Research Ltd.
VIA Rail Canada

FACILITIES

Moorelands Camp
Trent University
St. Albans Boys and Girls Club

FOOTNOTES

1. The global cooling that gave rise to the ice ages was likely the result of the concentration of land masses near the northern polar region and the southern polar location of Antarctica. The cycles of glacier advance and retreat appear to be caused by cyclic variations in the geometric relationship between the earth's orbit and the sun. We now enjoy a period of relatively warm global climate called an interglacial. Over the last two million years there have been other interglacials and it is likely that the climate in the Canadian Arctic today is not greatly different from earlier warm periods.

2. It is worth noting that over the last 6000 years the orbital geometry of the earth has been changing to the point where global cooling should again initiate ice sheet growth. The cooling required for a major glaciation is a slow process occurring over thousands of years. Some scientists feel that the 'greenhouse effect', caused by the increase in carbon dioxide and other gasses produced by modern technology, will cause enough warming of the earth's atmosphere to preclude another major glacial period. Given sufficient warming it would be possible for the arctic climate to retreat northward, leaving temperate climate conditions and a well forested Kazan valley in its wake.

3. Gyrfalcon (*Falco rusticolus*), willow ptarmigan (*Lagopus lagopus*), rock ptarmigan (*Lagopus mutus*), snowy owl (*Nyctea scandiaca*), and common raven (*Corvus corax*).

4. The same principle of heat transfer is the reason it takes a cake longer to bake than cookies - the heat of the oven has farther to travel to penetrate to the centre of the larger cake.

5. The first Atlas of Breeding Birds was produced in Britain from data collected between 1968 and 1972 inclusively. This project was so successful that the concept spread throughout Europe, to Australia, and even into some African countries. At present in North America there are approximately 40 atlas projects at the state or provincial level, including the landmark Ontario atlas published in 1987.

6. In comparison, small mammals tend to be nocturnal and must be trapped and often killed to be positively identified. Such procedures can be difficult to implement on a wide-scale basis.

7. World Wildlife Fund, 1988.

8. *Inuksuk* (plural *inuksuit*) is an Inuit word meaning “likeness of man”.

9. Knud Rasmussen, 1930. Observations on the Intellectual Culture of the Caribou Eskimos. *Report of the Fifth Thule Expedition 1921-24*, Volume 7 No. 2. Nordisk Forlag, Copenhagen. p 42.

10. see Burch, E., Jr. “Muskox and Man in the Central Canadian Subarctic 1689-1974.” *Arctic* 30 (1977): pp 135-154

11. Ernest S. Burch, Jr. “The Caribou Inuit” in *Native Peoples, the Canadian Experience* edited by R. Bruce Morrison and C. Roderick Wilson. McClelland and Stewart, Toronto. p. 129.

12. Arthur J. Ray, 1990. *The Canadian Fur Trade in the Industrial Age.* University of Toronto Press, Toronto. p. 95

13. Jean Gabus, 1947. *Iglous, vie des Esquimaux-caribou.* Editions Victor Attinger, Neuchatel. p. 45.

14. Samuel Hearne, 1958. *A Journey to the Northern Ocean*, edited by Richard Glover. Macmillan, Toronto. p. 23

15. Alphonse Gasté, letter to Bishop Grandin, July 15, 1869, translated and edited by G-M Roussilière. *Eskimo* 57 (1960), p. 10-11

16. J. Burr Tyrrell, “A Second Expedition through the Barren Lands of Northern Canada”. *Geographical Journal* 6 (1895), p. 442

17. For a fuller discussion of Indian-Inuit relations, see Janes, R. “Indian and Eskimo contact in Southern Keewatin: an ethnohistorical approach”, *Ethnohistory* 20 (1973): 39-54; and Smith and Burch under suggested readings.

18. Rasmussen *op. cit.* p. 35

19. Birket-Smith, K. “The Caribou Eskimos; Material and Social Life and Their Cultural Position. Descriptive Part.” *Report of the Fifth Thule Expedition 1921-24*, Vol. 5, Part I. Nordisk Forlag, Copenhagen, 1929. p. 262.

20. For documentation of these routes within living memory, see: Milton Freeman Associates, Ltd. *Inuit Land Use and Occupancy Project*, (3 Vols.) Dept. of Indian and Northern Affairs, Ottawa, 1976.

21. Samuel Hearne, *A Journey to the Northern Ocean.* Macmillan and Co., Toronto, 1958 pp 203-204

NOTES ON CONTRIBUTORS

David F. Pelly is an avid barrenlands canoeist, popular historian, and freelance writer. Beginning with his first book, *Expedition*, he has been widely published on the Canadian north in a variety of magazines, and subsequent books, including *Qikaaluktut: Images of Inuit Life*, a collection of illustrated stories depicting the lifestyle of the inland Inuit. The Kazan River expedition was David's idea.

Christopher C. Hanks has lived in Yellowknife for the past nine years, working as the Subarctic History Advisor for the Canadian Parks Service and previously as the Subarctic Archaeologist at the Prince of Wales Northern Heritage Centre. He has carried out ethnographic, historic, and archaeological field work across the subarctic, the results of which have been published in several scientific journals.

Glen MacDonald is an Associate Professor of Geography at McMaster University in Hamilton, Ontario. The results of his palaeobotanical studies in western and arctic Canada have been widely published. His current research includes work on the migration of jack pine, the 'ice free corridor', and the movement of the treeline.

Jane Claricoates is with The Wildfowl and Wetlands Trust, in Slimbridge, England. Her past research includes work on acid peat systems in Britain; and botanical, bird, and mammal fieldwork in Spitsbergen and west Greenland.

Judith Kennedy is employed by the Canadian Wildlife Service in Ottawa, Ontario. Previously, she was assistant co-ordinator of the Ontario breeding bird atlas and co-ordinator of the Maritime breeding bird atlas. Judith also has extensive experience as a nature interpreter and guide on wilderness canoe trips.

Andrew Stewart is a research associate at the Royal Ontario Museum in Toronto, Ontario. He is using the archaeological data derived from the Canadian Arctic Expedition for his doctoral dissertation at the University of California - Santa Barbara.

Mary McCreadie, formerly with the NWT Women's Secretariat in Yellowknife, NWT, is now freelancing as a consultant on women's issues. She is volume editor of a forthcoming book on NWT wilderness canoe routes.

PERSONNEL – STAFF

Expedition Leader	David F. Pelly (Canada)
Group Leader	Mary McCreadie (Canada)
Group Leader (Official Photographer)	Marc Côté (Canada)
Group Leader (Expedition Physican)	Mike Whittier (Canada)
Chief Scientist	Christopher C. Hanks (Canada)
Archaeologist	Andrew Stewart (Canada)
Ecologist	Jane Claricoates (England)
Biologist	Judith Kennedy (Canada)
Training/Basecamp	James Raffan (Canada)
Baker Lake Liaison	Gail C. Simmons (Canada)

VENTURERS

Richard Furhoff	Australia
Kassie Heath	Australia
Sonia Mellor	Australia
George Whitehead	Australia
Annie Roberts	Bahamas
Betty Ann Betsedea	Canada
Bruno Drolet	Canada
Almon MacNeil	Canada
Leslie Mack-Mumford	Canada
Simon Cremer	England
Ian Foulger	England
Jeremy Tate	England
Hilli Woodward	England
Choi Siu Ping	Hong Kong
Paul Clements	Jersey
Osama Abdeen	Jordan
Abdul Hasnie	Pakistan
Richard Wilson	Scotland
Boy Mow Chau	Singapore
Eddy Chong	Singapore
Angela Haas	United States
Scott McNamee	United States
Ashley Wooten	United States

INDEX

PHOTO / ILLUSTRATION CREDITS

P. 10, 92, 100
J.B. Tyrrell Collection, Thos Fisher Library, University of Toronto.

P. 14, color section,
Mark Côté

P. 29, 34, 42, 44, 64, 66, 103, 106, 116, 118, 121
Kassie Heath

P. 112
Andrew Stewart

P. 123
Betty Ann Betsedea